Philosophy in a Technological World

Also available from Bloomsbury

Advances in Experimental Philosophy of Science, edited by Daniel A. Wilkenfeld and Richard Samuels
Contemporary Philosophy and Social Science, edited by Michiru Nagatsu and Attilia Ruzzene
Nothingness and the Meaning of Life, by Nicholas Waghorn
Philosophy in a Meaningless Life, by James Tartaglia

Philosophy in a Technological World

Gods and Titans

James Tartaglia

BLOOMSBURY ACADEMIC
LONDON • NEW YORK • OXFORD • NEW DELHI • SYDNEY

BLOOMSBURY ACADEMIC
Bloomsbury Publishing Plc
50 Bedford Square, London, WC1B 3DP, UK
1385 Broadway, New York, NY 10018, USA
29 Earlsfort Terrace, Dublin 2, Ireland

BLOOMSBURY, BLOOMSBURY ACADEMIC and the Diana logo are trademarks of
Bloomsbury Publishing Plc

First published in Great Britain 2020
This paperback edition published in 2022

Copyright © James Tartaglia, 2020

James Tartaglia has asserted his right under the Copyright, Designs and Patents Act, 1988,
to be identified as Author of this work.

For legal purposes the Acknowledgements on p. vii constitute an extension of this
copyright page.

Cover design: Charlotte James
Cover image: Detail from 'Space Time' by Santiago Ribeiro. Oil on canvas.
© 2001 Courtesy of the Artist.

All rights reserved. No part of this publication may be reproduced or transmitted in
any form or by any means, electronic or mechanical, including photocopying,
recording, or any information storage or retrieval system, without prior
permission in writing from the publishers.

Bloomsbury Publishing Plc does not have any control over, or responsibility for, any
third-party websites referred to or in this book. All internet addresses given in this
book were correct at the time of going to press. The author and publisher regret
any inconvenience caused if addresses have changed or sites have ceased to
exist, but can accept no responsibility for any such changes.

A catalogue record for this book is available from the British Library.

Library of Congress Cataloging-in-Publication Data

Names: Tartaglia, James, 1973- author.
Title: Philosophy in a technological world: Gods and titans / James Tartaglia.
Description: London, UK; New York, NY, USA: Bloomsbury Academic, 2020. |
Includes bibliographical references and index.
Identifiers: LCCN 2020026795 (print) | LCCN 2020026796 (ebook) |
ISBN 9781350070103 (hardback) | ISBN 9781350185012 (paperback) |
ISBN 9781350070127 (epub) | ISBN 9781350070110 (ebook)
Subjects: LCSH: Philosophy, Modern–21st century. | Technology. |
Philosophy, Modern. | Materialism.
Classification: LCC B805.T36 2020 (print) | LCC B805 (ebook) | DDC 190–dc23
LC record available at https://lccn.loc.gov/2020026795
LC ebook record available at https://lccn.loc.gov/2020026796

ISBN: HB: 978-1-3500-7010-3
PB: 978-1-3501-8501-2
ePDF: 978-1-3500-7011-0
eBook: 978-1-3500-7012-7

Typeset by RefineCatch Limited, Bungay, Suffolk

To find out more about our authors and books visit www.bloomsbury.com
and sign up for our newsletters.

For the calypsonians

Contents

Preface		ix
A Note about the Endnotes		x
Introduction: Disturbed by the Thought of Philosophy		1
1	**A World Without Philosophy**	13
	§1. Imagining the world without philosophy	13
	§2. Gods and Titans	19
	§3. The Problem of Ceaseless Technological Advance	25
2	**The Materialist Philosophy**	31
	§1. The Metaphysical Issue	31
	§2. Cold War Metaphysics	33
	§3. Disenfranchisement	37
	§4. Scientism	40
	§5. Weariness	42
3	**When Philosophy Lost Its Mind**	47
	§1. Is Materialism Plausible?	47
	§2. Two Paths to the Standard Picture	49
	§3. The Lacuna	52
	§4. The Let-Down	60
	§5. The Loop	68
4	**A New Idealism**	71
	§1. Seeing the World in a New Light	71
	§2. An Outline of Idealism, and of its Consequences	75
	§3. Five Arguments for Idealism	82
5	**Technoparalysis**	95
	§1. A Positive Proposal	95
	§2. Inevitability and Desire	97
	§3. The Need for Greater Philosophical Awareness	106
	§4. Better Angels and *Homo rapiens*	110

6	**Freedom**	119
	§1. An Anomaly	119
	§2. Constraints and Restraints on Freedom	120
	§3. The Idealist Solution	124
	§4. Astrology and the Metaphysics of Desire	135
7	**Soul**	141
	§1. Making Souls	141
	§2. Ego Death and Substance	148
	§3. Video Games	156
8	**Truth**	161
	§1. The Value of Balance and Post-Truth Culture	161
	§2. Philosophical Scepticism about Truth	162
	§3. Truth First	166
	§4. Philosophy's Traditional Obligation	168
	§5. A Utopia	172
	§6. Take Stock	178
Notes		181
Bibliography		197
Index		207

Preface

I discovered calypso while I was writing this book. It originated in Trinidad in the nineteenth century and was first recorded in 1912, five years before jazz. My 'calypso craze' happened rather late – I think most of the world had it in the 1950s. But it did at least happen and I cannot thank my friend Allan Gonzalez Estrada enough for that. Thanks to Allan, I had Tiger, Caresser, Radio, Growler, Melody, Kitchener, Sparrow, Fighter, Walter Ferguson and Cro Cro to help me through the writing process; he even got Ferguson to record a personal message to me. Calypso contains plenty of philosophy and some of the calypsonians had their own concerns about technology: Growling Tiger's 'Atomic Energy Calypso', Mighty Spoiler's 'Mad Scientist' and Walter Ferguson's 'Computer' are all good examples – but note that I was not influenced. There is even a fragment of calypso lodged firmly within the mainstream of academic philosophy. For every philosopher knows half a line from Roaring Lion's song, 'Ugly Woman', since it inspired the title of W.V.O. Quine's *From a Logical Point of View*: 'Therefore, from a logical point of view', sang Lion, 'always marry a woman uglier than you.' I failed to heed that advice but it turned out just fine. It was during a memorable late-night barbeque with my wife, Zo Hoida, that she came up with my favourite line in this book (at the end of Section 3, Chapter 3); I only had to clean it up a little bit. My best friend, Commander Steve Forge of the Royal Navy, inspired the discussion of Steven Pinker and John Gray (Section 4, Chapter 5). Raymond Tallis's questions about the penultimate draft led me to the missing links in the argument of the book, which transformed the final product into what I always wanted it to be; Kelly Harmon's questions and reservations were very helpful too. The draft they read had already been seriously improved by Stephen Leach's judgement and learning. Simply knowing some people helps me do what I do, so a big shout out to: Adam Balmer, Kieran Brayford, Alex Brecker, Tom Clark, Tim Crane, Jacob Fox, Philip Goff, John Horton, Adam Kimberley, Alan Malachowski, Jennifer McCarthy, Martin Müller, Sila Özdemir, Bjørn Ramberg, G.A.J. Rogers, Artur Szutta, Natasza Szutta, Steve Tromans, J.J. Valberg and Emil Visnovsky.

James Tartaglia, Royal Sutton Coldfield, 29 February 2020

A Note about the Endnotes

As in *Philosophy in a Meaningless Life*, I have referenced endnotes in two different ways: superscript[1] when I have something additional to say and subscript[1] when I am mainly just providing bibliographic information or making connections to other texts. That way you never have to turn to the end of the book to find nothing but a page number.

Introduction: Disturbed by the Thought of Philosophy

The titan Prometheus defied the gods to give human beings the gift of fire. It was a dangerous magic with the power to hurt, destroy and transform, which gave us warmth, light and comfort. Prometheus's magic fascinated us and still does. In gratitude for his gift, we devised a philosophy in his honour named 'materialism'.

Physical science has allowed us to find innumerable technological applications for Prometheus's gift, as has precision engineering and many other endeavours that are not physics. The current manifestation of the materialist philosophy maintains that physical science tells us the ultimate truth about reality. Materialism dominates our world as secular common sense when it comes to philosophical matters. Outside of academic philosophy, it is very widely supposed that everything that exists ultimately consists in the particles and forces described by physics, and that existence itself began with the Big Bang. Where there is disagreement with this picture, it is most likely to have religious inspiration. Materialism is the respectable default within today's academic philosophy too, in that it is the only philosophy of the nature of reality which can be presupposed without argument in a journal article, with any suggestion of difficulties for materialism being automatically deemed worthy of interest. Materialism, as Hilary Putnam observed in the 1980s, and as is no less accurate today, is the only metaphysical philosophy with 'contemporary "clout"'.[1]

Given this situation, you might be forgiven for expecting scientists, and especially physicists, to love philosophy. But although many working scientists do indeed, the message being sent out about philosophy by the most prominent public spokespeople of science in their popular books, broadcasts, interviews, and – increasingly – social media, is so thoroughly negative as to be puzzling.

The best-known example is:

Philosophy is dead.

Stephen Hawking said that – the greatest hero of recent science, who overcame debilitating illness to revolutionise our understanding of black holes.[2] Why did he say it? Because he did not think philosophy had kept up with the latest developments in physics. But why should it? Division of labour requires that chemists focus on chemistry, accountants on accountancy, and philosophers on philosophy. Imagine there being some future breakthrough in biology which is relevant to physics, but which physicists ignore. It might then be appropriate for a prominent physicist to chastise their own profession for not keeping up with developments in biology, but not for a biologist to declare that physics is dead. Hawking's statement suggests that philosophy is unique among academic disciplines, other than physics, in being something physicists know best about; and materialist philosophy does say something in that vicinity. Despite what Hawking believed, however, many philosophers have gone to extraordinary lengths to keep up with contemporary science, and particularly physics.[3] If he was just badly informed, and a philosophical focus on physics is what matters most, then what he should really have said is that philosophy has never been healthier.

Astrophysicist Neil deGrasse Tyson agrees wholeheartedly with Hawking's statement. During a knockabout interview in which a variety of philosophical questions had been discussed, and practically nothing else – for example, Tyson speculates about whether the universe is a computer simulation, and if it was, whether he would choose to stay within it (he would not) – the interviewer mentions that he studied philosophy at university. Tyson immediately butts in:

That can really mess you up![4]

They all proceed to laugh at how 'futile' philosophy is, with the interviewer calling it 'a fat load of crap', and Tyson correcting the suggestion that it is good for comedy, since 'you need people to laugh at your ridiculous questions'. He tells the one about the scientist and the philosopher crossing the road – the scientist says to the philosopher:

Look, I got all this world of unknown out there. I'm moving on. I'm leaving you behind. You can't even cross the street because you are distracted by what you are sure are deep questions you've asked yourself. I don't have the time for that.

Here we see to the heart of his sentiment: keep moving on and never let philosophical qualms get in your way. The story might have ended with the scientist being run over. According to Tyson's history, however, philosophy once helped science to move on before it was left behind by the scientific revolutions of the 1920s, after which it became a hindrance. And yet ever since the late 1950s, when Tyson was born, philosophy has embraced materialism.

The particle physicist Brian Cox has said:

I don't 'do philosophy' in the same way that I don't 'do homeopathy'.[5]

Homeopathy has been scientifically discredited – after a substance has been diluted to the prescribed levels it is ineffective. But since there is no testable claim that all philosophy relies upon, this makes you wonder what Cox thinks philosophy is. The answer both he and Tyson are working with, I suggest, is that philosophy is anything you *cannot* scientifically test. When discussing the origins of reality, for instance, Cox says there is some speculation among scientists that the Big Bang may not have been the beginning of the universe, since the universe may have always existed. Then he says: 'Whether or not this would be a satisfying answer is up to you. I'd be comfortable with it.'[6] What he must mean by a 'satisfying answer' is a philosophically satisfying answer, since he cannot be allowing that the science could ever be a matter of personal satisfaction and comfort. That would explain the comparison to homeopathy. The believer in homeopathy ignores science because they find homeopathy satisfying, just as the philosopher ignores science because they find certain philosophical views satisfying. Not all reasons are scientific reasons, however, and philosophy is famous for its focus on arguments and reasons. I expect Cox thinks long and hard about non-scientific reasons when deciding how to vote, for example.

According to biologist Richard Dawkins, the physicist Lawrence Krauss's book *A Universe from Nothing: Why there is Something rather than Nothing*, may deal the deadliest blow to religion since Darwin's *On the Origin of Species*.[7] From 'staggeringly beautiful experimental observations', as Krauss puts it, he discovered that 'getting something from nothing is not a problem'. Since the idea of observing nothing is very puzzling, Krauss realises he needs to say something about philosophy at the start of the book:

I have learned that, when discussing this question in public forums, nothing upsets the philosophers and theologians who disagree with me more than the notion that I, as a scientist, do not truly understand 'nothing'.[8]

He then presents a philosophical argument:

For surely 'nothing' is every bit as physical as 'something', especially if it is to be defined as the 'absence of something'. It then behoves us to understand precisely the physical nature of both these quantities. And without science, any definition is just words.

If 'nothing' has a physical nature then 'nothing' is something with a physical nature. If that is 'just words' then what Krauss said is too, since there is no science behind it. It seems to me that if you want to avoid philosophy, you cannot do so by making ill-considered philosophical claims while showing disdain for philosophy, any more than you can avoid a game by playing it badly and saying you do not like it. If Krauss is uninterested in the philosophical question – which I find hard to believe given the area he went into – then why do everything possible to create the impression that this is the question he has answered, while letting Dawkins talk about religion at the end of the book? He could have easily avoided the issue by claiming to have shown how the universe sprang from a 'minimal something', for instance. That he did not reveals the influence of the materialist philosophy: he was driven by the thought that physics cannot be allowed to leave such a large and obvious mystery unaddressed. When David Albert, a philosopher of science with a solid background in theoretical physics, reviewed the book to point out that Krauss's understanding of 'nothing' obviously has a very strong bearing on the philosophical issue, Krauss called him a 'moronic philosopher'. He later apologised, while restating his claim about definitions needing science, and saying that philosophers who cannot accept this do not interest him, the others being okay.[9]

Not all public physicists are quite so negative about philosophy, but the need to pass judgement on it does seem to obsess them. Going back in time a little to the 1990s, we find a chapter of Nobel laureate Steven Weinberg's book, *Dreams of a Final Theory*, entitled 'Against Philosophy', in which he does at least find a positive use for it: for dismantling philosophical preconceptions which get in the way of science. The ultimate goal of this task would presumably be for nobody to think philosophically about science anymore.[10] But if we go back even further, we see what a recent phenomenon this all is. Consider the most acclaimed publicly engaged physicist of all, Albert Einstein:

> A knowledge of the historic and philosophical background gives that kind of independence from prejudices of his generation from which most scientists are suffering. This independence created by philosophical insight is – in my opinion – the mark of distinction between a mere artisan or specialist and a real seeker after truth.[11]

Einstein was always interested in philosophy and made it his business to converse with philosophers, fully aware that his work had many possible philosophical implications which needed to be worked through and rationally debated.

But perhaps we should put aside physics to see if a rosier view of philosophy is presented by spokespeople for other sciences. Psychologist Steven Pinker delivers just that throughout his various works, but even here there is an important caveat:

> Today most philosophers (at least in the analytic or Anglo-American tradition) subscribe to naturalism, the position that 'reality' is exhausted by nature, containing nothing 'supernatural,' and that the scientific method should be used to investigate all areas of reality, including the 'human spirit'.[12]

He is not wrong. But what he means by 'naturalism' is basically materialism, since as many of his writings make clear (I discuss Pinker in Chapter 5), he would count any philosophical view of reality with a non-physical component, such as idealism or dualism, as committed to the supernatural. So, what we see here is some reassurance being offered to readers: 'don't worry about philosophy because it is respectably materialist these days.' We finally see some appreciation being shown for philosophy's turn to materialism. It is no longer poisoned by archaic, anti-scientific nonsense, because it recently made the right choice in a debate that existed in ancient Greece.

Pinker is quite the exception, however. Biologist Edward O. Wilson, despite being very positive about the humanities in general, has this to say about philosophy:

> I like to say that most of philosophy, which is a declining and highly endangered academic species, incidentally, consists of failed models of how the brain works.[13]

He said this while trying to justify his own parallel move to that of Krauss, namely to show that science can solve a traditional philosophical problem, this time the meaning of life.[14] The statement itself is very odd, however. Since models of how the brain works have not loomed large in the history of philosophy, what he must mean is that when philosophers thought they were theorising about reality, knowledge and morality, for example, they were really, unwittingly, trying to understand the workings of human brains and getting it wrong. Brains are part of reality, however. So, if you need to know about the workings of brains to know about reality, then you must need to know about the workings before you can find out about those workings, which is impossible.

Richard Dawkins says that, 'At its best, philosophy can aid understanding. At its worst, its jargon supplies a handy toolkit for charlatans to bamboozle the innocent'.[15] That is fair comment, but it should be added that dressing up philosophical opinion as scientific fact provides a particularly powerful toolkit for bamboozling the innocent at present. But even Dawkins, thoroughly embroiled in philosophy as he is, has not been able to resist the urge to make a blanket denouncement:

Philosophers' historic failure to anticipate Darwin is a severe indictment of philosophy.[16]

And yet the concept of natural selection has a history stretching back to Empedocles.[17] There were over two thousand years of philosophical anticipations, but Darwin was a biologist, so philosophers could not have been expected to anticipate his biology. They were quick to read philosophical significance into it, however, with Nietzsche being a good example.

All of the scientists mentioned above write books which make philosophical claims. Some are full to the brim with philosophy, as you might expect from titles like Wilson's *The Meaning of Human Existence*, Krauss's *A Universe from Nothing*, Hawking and Mlodinow's *The Grand Design* and Dawkins' *The God Delusion*. This makes their denouncements of philosophy particularly puzzling. Academic philosophy has never been more materialist, and this builds deference to science into the discipline in a manner not encountered in other areas of the humanities. Similar philosophical commitments are widespread outside the academy, except where religious beliefs prevent them. So, the atmosphere has never been more conducive to reading philosophical significance into scientific knowledge, which seems to be exactly what these scientists want to do. And yet they denounce philosophy and, presumably, do not think they are making philosophical statements.

What we are seeing, I think, is the influence of the materialist philosophy. It only became established in the English-speaking world in the middle of the twentieth century and has now reached maturity. The present generation of public scientists have thoroughly absorbed it, just like the rest of us, and it says that science is the best route to truth. So as scientists, they have been encouraged to feel they are the only ones with a right to talk about philosophy. This provides a personal motivation for their attacks on the non-scientist philosophers, who, while they still exist, pose a disconcerting threat to their freedom in this regard. But I think it goes deeper than that.

The deeper explanation is that to acknowledge that there is a legitimate area of concern called 'philosophy' gets too close for comfort to recognising that their materialist convictions are philosophical. For if materialism is a philosophical view, and it might be a false one, then there might be philosophical questions about reality which science cannot address – big ones, like why the universe exists or the meaning of life. A wholesale denouncement of philosophy removes any potential for having your materialist convictions challenged. So, I think the scientists most inclined to denounce philosophy are those with the deepest, most uncritical love of it – but only philosophy of one particular kind. Since materialism is not science, the thought of philosophy disturbs them. When they denounce philosophy, they yield to the demands of their own.

The reason I do not take these denouncements lightly, as many might think I should, is that they seem to me symptomatic of a wider situation in which philosophical reflection is coming to seem less and less important, while the power of science and technology to change the basic conditions of human life are rapidly increasing. I think this is a bad combination, because as more technological transformations of the conditions in which people live their lives become possible, the more we should be philosophically reflecting on which of the transformations we want to enact. The more we *can* do, the more we should reflect on what we *want* to do – where the 'we' who should reflect is 'as many people as possible'. It seems to me that materialist philosophy actively discourages the kind of widespread philosophical development we need in order to keep pace with technological development and thereby allow it to improve our lives in a rationally constrained and popularly mandated framework. I also think materialism is false, which is an excellent traditional reason for not believing something and thereby allowing it to alter your behaviour.

Hawking said 'philosophy is dead', but look at the kind of concerns he expressed in *A Brief History of Time*, when connecting Darwin's theory of evolution with the quest for a single, unified theory in physics:

> It has certainly been true in the past that what we call intelligence and scientific discovery have conveyed a survival advantage. It is not so clear that this is still the case: our scientific discoveries may well destroy us all, and even if they don't, a complete unified theory may not make much difference to our chances of survival. However, provided the universe has evolved in a regular way, we might expect that the reasoning abilities that natural selection has given us would be valid also in our search for a complete unified theory, and so would not lead us to the wrong conclusions. Because the partial theories that we already have are sufficient to make accurate predictions in all but the most extreme situations, the search for the ultimate theory of the universe seems difficult to justify on practical grounds. (It is worth noting, though, that similar arguments could have been used against both relativity and quantum mechanics, and these theories have given us both nuclear energy and the microelectronics revolution!) The discovery of a complete unified theory, therefore, may not aid the survival of our species. It may not even affect our lifestyle. But ever since the dawn of civilization, people have not been content to see events as unconnected and inexplicable. They have craved an understanding of the underlying order in the world. Today we still yearn to know why we are here and where we came from. Humanity's deepest desire for knowledge is justification enough for our continuing quest. And our goal is nothing less than a complete description of the universe we live in.[18]

He says 'our scientific discoveries may well destroy us all'. Like the rest of us, it does not keep me up at night; news about nuclear proliferation occasionally invades your consciousness then floats away again – most of us have never known any different. But there is clearly an important debate to be had about whether we actually want physicists to press on towards their final theory, if doomsday is the risk involved. All that craving to understand which he talks about might well be considered a very minor factor to consider if a debate were to transpire among all the relevant stakeholders. Perhaps the yearning is more philosophical than scientific. Perhaps the latter is largely confined to the scientists doing the research, and most people, unable to understand it properly anyway, only really care about the technological consequences and how they change their lives. Would they be wrong to think that? Would asking for some restraint show that people do not know what is good for them, or are showing insufficient gratitude to science?

The great jazz saxophonist Sonny Rollins has said:

> Everything about technology, folks, is not good. Hate to tell you, folks, but it's not all good.[19]

The way he says it ('Hate to tell you, folks') is a reminder that these days, it needs to be said again, and again, and again. Natural positivity and optimism about life, combined with one-sided views tirelessly promoted by those with vested interests, make it all too easy to forget. The occasion for Rollins' comment was provided by a spoof article purporting to be a confessional piece he had written, which was widely circulated on the internet (it 'went viral'). The article portrayed him as someone who hates jazz and believes himself to have 'wasted [his] life' – infantile, but no big deal. But deep down we all know that internet technology is 'not all good' and that this is sometimes a really big deal. We know this because it changed our lives and we know what our lives are like now. Hours spent trying to get your computer working again are not good hours; an inbox full of hundreds of emails will rarely make your spirits soar; attempts to trick you popping up onto your screen are annoying; worrying about an illustrated encyclopaedia of depravity and malice at the end of your children's fingertips is a problem we never used to have. Quite possibly it made things better than they were before, although I remember no beautiful emergence from a cocoon during those years when I, like the rest of my generation, started using the internet. There was certainly no widespread, all-consuming debate about whether we should transform our lives in the widespread, all-consuming way we have. Now far greater transformations are envisaged, and they are engineering projects, not propositions to consider. A more balanced, rational and pro-active attitude to technology needs to develop among its consumers, and a more balanced and rational attitude needs to develop among its producers.

This is a standard theme in the philosophy of technology. Hans Jonas advocated a new, more consensual ethics developing around technological development, and many others have followed suit.[20] But philosophy of technology remains a minor area within the wider academic discipline. Langdon Winner felt justified in saying, back in 1986, that 'the most accurate observation to be made about the philosophy of technology is that there really isn't one'.[21] Whatever truth there was in that – it depends on his view that there was 'little of enduring substance' in the thousands of books and articles he surveyed – things are changing. Increasing numbers of philosophers now put their minds to the question Jonas and Winner prioritized, namely how to establish limits in a world in which science and technology are continually expanding the scope of what it is possible for us to do. To make these efforts practical requires being adequately informed about the working practices involved in technological development, as well as the funding decisions that get them started, and much collaboration now takes place with the interdisciplinary field of Technology and Science Studies.[22] A discipline of Engineering Ethics has been established (Langdon found such considerations to be completely alien to engineers in 1986[23]), as well as Computer Ethics, Nanoethics, and various other subdisciplines devoted to specific developing technologies. There is a *Centre for the Study of Existential Risk* at Cambridge University, co-founded by a philosopher (Huw Price), an engineer (Jaan Tallinn) and a scientist (Martin Rees), as well as philosopher Nick Bostrom's *Future of Humanity Institute* at Oxford University.

Such developments are to be welcomed without reservation, but they face a very serious uphill struggle. As a member of the public, one who actively listens out for this kind of thing, I have noticed no sign of widespread, all-consuming debates starting to materialize over the particularly dramatic new technologies currently envisaged. When I hear politicians mention artificial intelligence, they are talking about the economic benefits which they promise not to allow my country to miss out on. When I hear about major developments to neural implantation technology, this is because it holds out the prospect of curing Parkinson's Disease – an exceptionally powerful pro, now what about the cons? Concerns are always dismissed, typically seen as a source of amusement or sign of ignorance; details and arguments are always absent. I do not feel I am giving informed consent to developments that will fundamentally alter the future of human life, and I very much doubt that this is because I am a philosopher, rather than an airline pilot, nurse, architect or builder. It seems to me that something very dramatic is going to have to change before this process of radical change can be considered remotely democratic.

I am not silly enough to think that academic philosophy will lead the way. Plato tried some direct action and regretted it on his return from Syracuse. Academic philosophy exerts its influence more indirectly, as vague contours of new ways of thinking gradually catch on to alter behaviour in the long run. What I do think, however, is that 'philosophy' in a more general sense, one which explains why there is an academic discipline, is something that could spread far beyond the academy to make a decisive difference to how human beings develop in a technological world. It is well-placed to do that because it is neither science nor religion, but can rationally reflect on both. I defended this conception of philosophy in my previous book, and I also published a paper defending it in the appropriate academic journal. It is a task very few philosophers undertake. My leading thought was that there must be some kind of subject-matter which explains why there should be a discipline which encompasses fields as seemingly diverse as metaphysics and ethics. No criticism of my conception of philosophy has ever been made, to my knowledge, so I feel justified in proceeding to work with it, as I shall do in this book. In a nutshell, it is that, 'Philosophy is the study of a range of related issues concerning knowledge, reality, and moral conduct, which traditionally centre on the question of life's meaning'; my use of the word conforms to that conception throughout.[24]

I think materialism is the main philosophical obstacle to philosophy (in my generalist sense) becoming a more widespread, self-conscious preoccupation which might benefit our approach to technological development. Materialism not only blurs the boundaries between science and philosophy, but works to actively discourage the notion of philosophy as a distinct field of interest. It also encourages apathy, in that it is liable to stand against an image of people as conscious free agents who determine their own future by independently thinking through the available options to try to make rational decisions in light of the truth. Materialism has this in common with another major current of twentieth century thinking to which it is instinctively opposed, namely the counter-Enlightenment currents of de-centring thought associated with the likes of Freud, Durkheim, Barthes and Derrida.[25] Materialism is similarly attracted to a picture of us as powerless pawns in a game played by nobody. Philosophy, however, is an assertion of rational autonomy, even when that autonomy is being used to deny itself. And for philosophy to spread, I think, it must draw on the best resources at its disposal, namely the natural interest of the traditional problems of philosophy, which need to be shown as relevant to the problems we face today.

As such, I shall be arguing against materialism, suggesting an alternative, and talking about some traditional problems in light of our contemporary situation. This book is not a specialized monograph on the philosophy of technology, just as my previous book, *Philosophy in a Meaningless Life*, was not a specialized monograph on the meaning of life. Once again, I aim to show the continuing relevance of the traditional problems of philosophy to matters outside the academy, while arguing for particular views on them within it, albeit in a manner that might be understood from without – attempting this balancing act is what my conception of philosophy, and of its value, requires. This time technology is my theme, and the more specific traditional problems are materialism and idealism, freedom, personal identity and truth.

How these topics fit into the overall argument of the book can be understood as follows. Materialist philosophy has exerted major historical influence over how we think about ourselves and our collective future. As the largely unreflected belief-system it has now become, it continues to shape the directions of our technological development, while encouraging us to think these directions are inevitable, that we have no freedom to do anything about it, that seeking truth is a specialist pursuit, and that our very identities are within the scope of technological development; that humanity itself is within that scope, in fact, and might even be worth replacing. Materialism has never been a rationally established philosophy, however, and for most of its history was embraced as a political agenda opposed to organized religion, with the technological advances of the twentieth century falsely seeming to vindicate it. Now it obstructs the kind of widespread public reflection which might break the current deadlock between resigned pessimism and blinkered optimism over the development of radical new technologies. In the poetry of Lucretius, which conveyed materialism from the ancient world to the modern, the myth of the war between gods and titans was interpreted in favour of materialism: in defiance of the gods, materialist philosophy sought technological aid from the titans to improve the human lot. But seeking this aid may result in horror, as we are reminded by Mary Shelley's novel *Frankenstein*, subtitled *The Modern Prometheus*. If technological development is to be driven by collective, rational debate, and the deadlock between pessimism and optimism broken, then we must find balance between gods and titans: between imagination and rational deliberation, and the power to enact our visions. I shall argue that a new, idealist philosophical understanding of ourselves, one which expands rather than challenges everyday understanding, encourages individual reflection, reasserts our freedom, and reflects the kind of lives we now live, would help us to find that balance.

I do not think that the materialist philosophy is an appropriate form of gratitude for Prometheus's gift. The appropriate gratitude is shown through the superlative status in modern life of scientists and inventors, alive and dead. Anyone inclined to doubt whether that status is appropriate, with their central heating, electric lights and flush toilets, smartphone in pocket and hospital within short driving distance, could do worse than to reflect on the doctor, philosopher and scientist Raymond Tallis's memorable question: 'How much of the history of human consciousness is a history of itching?'[26] Nevertheless, in the current intellectual climate, the argument of this book is liable to bring accusations of being anti-science, anti-technology, or just generally antimodern life. I am not sure it will help to say that I am none of these things; in my previous book, it was not altogether effective that I immediately and completely disassociated pessimism from my view that life is meaningless. But, for the record, I certainly do think that science tells us the truth about our physical environment; I just think such statements are open to competing philosophical interpretations, and that the interpretations are important. I am glad to have been born into an age of high technology and look forward to further developments within my lifetime, such as green technologies. I do not look wistfully back to a supposedly better bygone age; I think I would have been one of the peasants.

So, with the preliminaries over, let me tell you what will now follow. In Chapter 1, I will try to imagine what a world without philosophy might look like, before introducing the gods and titans myth, and what I shall call the 'problem of ceaseless technological advance'. In chapters 2 and 3, I will portray materialist philosophy, as well as philosophy itself to some extent, in a hopefully enlightening manner; simply the facts I will recount might be enough to have something of this effect. In Chapter 4, I will argue for an idealist philosophy which could replace it. Idealism has acquired a thoroughly bad name during the materialist era, as has metaphysics itself, but I think the version I defend fits in well with how we ordinarily think, as well as with those things we know – and know we do not know – which are most relevant to metaphysical assessment. In Chapter 5 I look at the deadlock between pessimism and optimism we currently face over concerns about technological development, and propose philosophical education as a way of breaking out of it. In Chapter 6, I argue that we are free, and explain how materialism and superstition have combined to create the false impression that this is impossible. In Chapter 7 I argue that – in a manner that requires some explanation given the contemporary connotations of the word – it makes sense to think of ourselves as souls. This understanding is forward-looking, not backward-looking, as I illustrate with a discussion of video games. Then I finish up in Chapter 8 with a reflection on the importance of truth, which too many powerful people seem to be forgetting about these days. This chapter includes a sketch of a utopia, with the final and very short section explaining its significance within the argument of the book.

1

A World Without Philosophy

§1. Imagining the world without philosophy

Without philosophy we would not be where we are now. In a trivial sense this is obvious, for we would not be where we are without tennis either, although there would be other games to build lives around. But without philosophy we might still be running with the animals. Natural philosophical curiosity about the ultimate nature of reality led to the cataloguing of fundamental elements and forces. It led to science. Perhaps the desire to master observable regularities could have taken us down that road in some other way, but speculation about what lay behind those regularities was the philosophical impulse which actually delivered. Similarly, curiosity about the meaning of life led us to the supernatural realm of religion. It is hard to imagine ritual practices developing without the gods they were meant to appease, and those gods were another explanatory principle arising from philosophical curiosity; another ultimate reality to stand behind the commonplace. Without religion, which is always philosophical at heart, we would have not been inspired, united and divided in the ways which led to our civilizations. We might not have had art, given that other animals do not and its earliest extant forms evidence religious inspiration. Could we have arrived somewhere like this without philosophy? Nobody knows. But it was philosophy that sought to draw a line between us and the other animals, and looking about us now, it seems to have succeeded.

Many would be prepared to celebrate the philosophical instinct, broadly construed, as an ancient impetus to all we were subsequently to achieve. But whatever might be thought about philosophy's historical role in the emergence of science and religion, both of these are clearly integral to our current situation. Without science and the technology it facilitates, and vice versa, we would be unceremoniously returned to the Stone Age. And if religious faith were to suddenly collapse among the large and growing majority of the world's population who have it, we have no idea what chaos might ensue; even militant atheists surely envisage a very gradual transition. It is hard to imagine our world without science and religion, then. But what would it be like without philosophy? Denouncements of philosophy have reached a crescendo in recent times, so let us try to imagine the situation being hoped for.

Universities would have one less kind of degree to offer, but would soon make up the numbers elsewhere. Academic philosophers would be out of a job, but they might be able to reapply their skills to science, mathematics, literature or history. The educated public would have one less field of interest to engage them, so lightly worn copies of *Thus Spoke Zarathustra* would have to part company from *The Picture of Dorian Gray*, *The God Delusion* and travel guides on casually erudite bookshelves.

But on second thoughts, universities would now have to be very careful what they were teaching, since philosophy has spread widely across the humanities and social sciences; anything written before the ban might be infected to some degree. Even natural scientists might sometimes feel the inclination to wax philosophical during the course of a one-hour lecture, so that would have to be curbed. And as for those casually erudite bookshelves, only the travel guides would really be safe. We might be able to save *Dorian Gray*, if it could be reworked to make it more boring, but not *The God Delusion*. There would have to be a widespread decimation of the popular science idiom which has filled a void left by religions and their philosophies in many people's lives. They usually contain plenty of philosophical speculation worked around reports from the scientific frontiers, since there is philosophical interest in alternative realities, the origins of the universe, human nature, eternal life, and so on. Perhaps we would travel more in a philosophy-free world, thereby broadening our minds in this other tried-and-tested fashion.

This may all be considered inconsequential on the grounds that philosophy is entertainment. We might be squeamish about tampering with great works of literature, theatre, film, music, poetry, paintings, conceptual art, etc., but you could hardly leave it intact in a world free of philosophy, lest it inspire the wrong thoughts. But since philosophy has no clear connection to what gets food on the table, and the extensive redactions to Shakespeare would no doubt be done expertly, let us reserve judgement for now and turn to something more practical.[1]

Politics is underpinned by philosophical commitment. Left- and right-wing politicians have different views about how society should be organized, in line with how they think we ought to live. In the background of their debates are left- or right-wing academics who use evidence to support action-plans for implementation. But motivating such plans are philosophical views, such as that states should try to maximise happiness, not interfere with individual liberties, and so on. These views emerged from reflection on what we should all desire. This is a paradigmatically philosophical notion because it alludes to a meaning of life, despite the fact that only the religious might instinctively make the connection nowadays. But the connection is there, because nothing could better secure the 'should' of *what we should all desire*, than the 'is' of a meaningful reality in which we have our place, such that we should all desire something because the nature of reality dictates that we should bring it about; just as the nature of reality dictates that the apple fall from the tree. Thoughts about what we should all desire become less philosophical, and more practical, when the scope of the 'should' reduces to 'should in order to bring about such and such results'. And less philosophical still when the 'we' reduces to 'we people of this nation' or 'we workers'. But given the root of all such thoughts, they remain very philosophical.

If philosophy were to go then politicians could no longer lay claim to deep commitments. But you might think this would be no bad thing on the grounds that real political action happens in debates about implementation. Still, it is hard to imagine political life without some kind of ideological split. We cannot reduce it to disagreements about implementation plans without deciding what we want to implement, but we can hardly agree on how we should live without engaging in exactly the kind of philosophical debate we are trying to eradicate.

Two possibilities suggest themselves. The softer one would be to leave the politicians with their desires for different outcomes, and eradicate reflection on the reasonableness of such desires. Politicians would desire what they desire now, draw up implementation strategies, and in democracies, the people would decide which strategies they desired. 'Wrong' would mean 'not part of the world which I, as a matter of brute fact, desire'. This is a compromise, however, because we would be leaving past philosophy embedded to be passed down the generations. The harder option would be to eradicate conflicting desires. The problem then arises of what we are trying to achieve, but perhaps neuroscience could settle the matter by looking inside our brains. They are physical things which evolved in similar circumstances, so it seems reasonable to suppose there is something they all want; maybe this has already been discovered.$_2$ Or maybe it is simply obvious that we all want food, shelter, security, happiness and eternally youthful life. Perhaps alternative desires are maladies to be treated, or are not really alternatives at all, because their neural reality is a convoluted desire for happiness. Perhaps we would not need politics at all once philosophy was gone. For surely intelligent machines would be better at working out the best implementation strategies for our utopia than the inevitably flawed, biologically implemented cognitive systems we are presently stuck with. Or we could genetically enhance politicians; perhaps we would only need one if the enhancement was good enough.

Another area of life that would be impacted is religion. Religious belief places the ordinary world we experience into a wider context of meaning, and typically holds that it is governed by something greater than, and concerned with, us. That is a very philosophical thing to believe, although the way it is often believed, namely as an unquestioned background belief, is not at all philosophical. In any case, since religious beliefs about gods imply philosophical views about reality, they would have to go. People could continue to engage in rituals and ways of life: they could enjoy the sense of community, calmly reflect in the meditative spaces of Christian churches and clap in front of Shinto shrines. But without the cultural institutions of belief, it would not be sustainable. Religion would expire with philosophy. Some will think we have now reached the most appealing aspect of this proposal.

But none of this is feasible. For even if we were able to eradicate philosophy from the world, leaving us with redacted Shakespeare, politics entirely concerned with implementation, and no religious faith, the philosophy would come right back at us – like a boomerang. Somebody would die, and stricken by grief, their loved ones would form the consoling thought that they intangibly lived on. A child would ask where the universe came from and talk of a Big Bang would simply push the question back a stage. An office worker, bored at her desk, would wonder what the point of it all was. A teenager would take psychoactive drugs and start thinking about consciousness. A scientist would wonder why her theory was so predictively successful. Questions would be raised about our political utopias and new ones would be dreamt up, better tailored to the ever-new living conditions we use technology to create. So, the proposal is not stable.

It seems it could only be made stable if, while making these changes, we also changed ourselves: by removing a philosophy gene or two. This is because we are philosophical beings. We *all* are, despite the fact that our philosophical natures are often suppressed before they can develop. We see this from how readily philosophical issues concern us when the social framework in which we live our lives is violently interrupted. And there is no more violent interruption than death; whether that of a loved one integral to your projects, or the imminent prospect of your own death placing you face to face with the termination of projects without which you do not know yourself. At these times people seek religious consolation, or otherwise engage in philosophical reflection. Since it is unrealistic to expect these instincts to ever be educated out of us, it is looking very much as if the eradication of philosophy is going to require neural alteration.

Now this all sounds very unattractive, of course, but the point of this exercise was obviously not to dissuade us from a concrete plan of action. Anti-philosophical sentiment is on the rise, and trying to follow through on that agenda has revealed its enormity. What we have seen is that philosophy is by no means confined to academic institutions, that losing this component of our lives is far from an obviously attractive prospect, and that there is no way we could enact the agenda in any case, short of plunging into a technological dystopia. Anti-philosophical sentiment is a real and increasingly significant phenomenon, which we have learnt to take lightly, but should not. This sentiment may be blind to its own direction of travel, but we should not.

In the real world, of course, those who denounce philosophy do not envisage a programme of eradication. If the denouncement is thoughtful enough to envisage anything at all, which is rare, then it is for philosophy to naturally fade from our horizons; although it is worth remembering that the eradication of religion from nation states has indeed been attempted by materialist ideologues. The scientists who denounce philosophy simply assume that science has all the answers, and are irritated by the existence of a discipline in which that could possibly be questioned – because of their commitment to materialist philosophy, as I said in the introduction. But philosophers who call for an end to philosophy – a real creed, which is something that has always intrigued me – think the matter through rather more carefully.

It was the nineteenth century positivist philosophy of Auguste Comte which provided the intellectual beginnings of the current anti-philosophy trend. Comte saw philosophy as a transitional stage on the road to science; as something the demise of which signals progress.$_3$ But it was Richard Rorty, in the late twentieth century, who developed the theme most thoroughly. Rorty immersed his life in philosophy to an extent few can ever have equalled and he would never have issued an unqualified call for the end of philosophy. He did, however, issue many qualified ones – because, in spite of his lifelong opposition to positivism in all its forms, he accepted the essentially positivist view that philosophy holds back human progress.$_4$ He tried to resolve this tension by proposing that philosophy be privatized. It became, for him, an inner, personal poetry, which insulated the intellectual elite by allowing them to gaze at the world in detached irony.$_5$ The irony, and hence philosophy, was never to be allowed into the public sphere, where it was the elite's moral duty to defend their views as vigorously as the non-ironic; as if the truth was on their side – which as the ironist realises, it never could be. In this vein, Rorty defended materialist 'truths': that everything is physical, that the mind does not exist (he pioneered this view), and so on. He would have been horrified at the prospect of extracting philosophy from literature and art, since that is exactly where he wanted it to be. The problem with philosophy, he thought, occurs only when people bring it into the public domain – when they think it raises real issues worthy of our collective attention.

Perhaps this provides a more realistic proposal to consider, then. Privatization might amount to much the same thing as eradication in the long-run, since philosophy makes claims about reality, and the interest of such claims depends on our ability to take them seriously; a supposedly true story loses much of its interest when you find out it never happened. Nevertheless, within a world of privatized philosophy, big philosophical ideas might still survive in our memories, cherished for their poetic content. Philosophy might live on as a shadow of its former self. I think Rorty thought it would be better that way.

Perhaps it could work to some extent. Perhaps people could learn to keep all but materialist philosophy to themselves. Books like Dawkins' *The Magic of Reality: How we know what's really true*, could be developed for even younger readers, such that the only philosophy we grew up with was trust in science.[6] A culture could develop in which whenever people found themselves raising philosophical questions in private, they knew better than to take them seriously, and so publicly confined themselves to demonstrable scientific truth. Religions could continue, but would keep out of the public domain – which is something Rorty was particularly keen on.[7] Visions of competing utopias might continue to animate political life, but nobody would feel the need to justify them with reasoned argument anymore, so long as there were more useful routes available to the same outcome. Philosophy would not have been eliminated, but rendered safe. Safe for what?

Science provides the root motivation for the anti-philosophy agenda. The motivation is that philosophy has been superseded by science, or soon will be. Apart from this essentially positivist motivation, which gets dressed up in a variety of ways, the only other consideration you regularly hear – which initially seems like an independent point, but is really just more of the same – is that philosophy never makes progress. (I shall explain why that is the wrong way to look at philosophy in the next chapter). This is just more of the same, because the intended contrast is with science. Thus, the conclusion we are meant to draw is that in a world of advanced science, philosophy is obsolete and counterproductive; unscientific, in short.

And yet if philosophy were to quit the public arena, developments in science would be divested of any wider rationality. Scientists would discover whatever they *could* discover, whether or not these discoveries were, on reflection, desirable. Philosophical views on how we want human life to develop and what we want to preserve of what we currently have would be absent. If even the scientists were not thinking these things through, the frontier of science would become a directionless lunge into the unknown. When the prospects of a radical technological transformation of our lives became imminent, people would worry, of course, but they would not question its inevitability. They might draw up contingency plans, but autonomous and unstoppable, science would move on regardless.

This is a concerning vision given that extremely recently, relative to the history of our species, science has unlocked awesome, unprecedented power which has brought us to the brink of total destruction on at least one occasion (the 1962 Cuban Missile Crisis), and which has proved incredibly difficult for our politicians to control ever since. Scientific knowledge is developing at an exponential rate, as scientists continually tell us, and the most vaunted breakthroughs on our horizon concern artificial intelligence, biotechnological manipulation of human beings, and more insight into the subatomic (e.g. 'The God Particle'); all areas which attract widespread public alarm because of their apparent dangers. On the face of it, philosophical reflection on what we want from science and where we can safely take it is now needed more than it has ever been. The notion of science as a disinterested quest for truth is immediately put into question by the very nature of the above-mentioned projects.

This looks like a definitive reason for keeping philosophy in our world. But we have now left the realms of the imagination. For when it comes to scientific and technological development, we no longer need to imagine a world without philosophy – we live in one. Philosophy is dead; or, more accurately, I think, philosophical reflection is dead because one philosophy is so deeply embedded. Perhaps it never lived in that sense, but it needs to now, at a time when the latest wave of scientific breakthroughs is feared but regarded as inevitable. It is as if they were giant asteroids on a long-term collision course with the Earth, rather than the imminent goals of ordinary people who brush their teeth before heading off to work in the morning. Some hope for a positive outcome, that the asteroids will change course. Others think they are not asteroids but manna from heaven, and they scorn contemplation of the alternative. And all the while the scientists and technologists labour away, racing to be part of the team that will fundamentally change human life in a yet to be determined manner. It is not for nothing that 'racing' has become the standard word to use in this context. Trying to find ways to safely manage and mitigate the negative effects of technologies regarded as an inevitable part of our future is the thoroughly unphilosophical order of our day.

§2. Gods and Titans

In Greek mythology, the Titanomachy is the war that took place between gods and titans. The gods won. With victory secured, they imprisoned most of the titans deep beneath the earth. The only exceptions were Atlas, the titan's war leader, who was reserved the special punishment of carrying the sky on his shoulders for all eternity, and the brothers Prometheus and Epimetheus, who had fought on the side of the gods. All the rest were imprisoned in Tartarus, and as Hesiod makes clear, the gods made every effort to ensure they would never escape:

> they sent them under
> The wide-pathed earth and bound them with cruel bonds–
> Having beaten them down despite their daring–
> As far under earth as the sky is above[8]

Hesiod adds that from their 'moldering place' of imprisonment, there is 'no way out for them. Poseidon set doors [o]f bronze in a wall that surrounds it.' Zeus also posted the monstrous Hecatonchires on guard-duty – just to be sure.

This was not the end of the troubles the titans were to cause the gods, however. For after emerging unscathed from the Titanomachy, Prometheus, the cleverest titan and the benefactor of humanity, tricked Zeus into accepting the fat and bones of animal sacrifice, leaving the most nourishing part for humans to eat. Zeus punished Prometheus by withholding fire from humans. Prometheus stole it and gave it to us. Zeus was furious, so he had the beguiling Pandora sent to Epimetheus as a gift, knowing she would wreak havoc among humans. Having been forewarned by his brother, however, Epimetheus declined. This was the final straw for Zeus, who ordered Prometheus to be chained to a pillar in the mountains where an eagle would gnaw at his liver all day, only for it to grow back during the night, ready for his next day of agony. Epimetheus now accepted Pandora, who opened her eponymous 'box' and thereby released all manner of evils upon humanity.[9]

More fallout from the Titanomachy was later to emerge in the form of the Gigantomachy: an uprising of giants against the gods, which was so similar to the first uprising that they have been conflated throughout our history.[10] According to Claudian, the second occurred because 'mother Earth [Gaia], jealous of the heavenly kingdoms and in pity for the ceaseless woes of the Titans [her children], filled all Tartarus with a monster brood'.[11] These monstrous giants then burst forth from the earth to challenge the gods for supremacy. Once more the gods emerged victorious, and once more they buried the vanquished deep beneath the earth.

In the *Sophist*, Plato compares the Gigantomachy to the philosophical dispute between materialists (a.k.a. physicalists[12]) and idealists:

Stranger What we shall see is something like a battle of gods and giants going on between them over their quarrel about reality.

Theaetetus How so?

Stranger One party is trying to drag everything down to earth out of heaven and the unseen, literally grasping rocks and trees in their hands, for they lay hold upon every stock and stone and strenuously affirm that real existence belongs only to that which can be handled and offers resistance to the touch. They define reality as the same thing as body, and as soon as one of the opposite party asserts that anything without a body is real, they are utterly contemptuous and will not listen to another word.

Theaetetus The people you describe are certainly a formidable crew. I have met quite a number of them before now.

Stranger Yes, and accordingly their adversaries are very wary in defending their position somewhere in the heights of the unseen, maintaining with all their force that true reality consists in certain intelligible and bodiless forms. In the clash of argument they shatter and pulverize those bodies which their opponents wield, and what those others allege to be true reality they call, not true being, but a sort of moving process of becoming. On this issue an interminable battle is always going on between the two camps.[13]

Plato sided with the gods / idealists in this 'interminable battle', but it was not long before materialists reinterpreted the analogy in their own favour. The Roman philosopher Lucretius made the decisive move in this passage:

I'll show you many things that will allay
Your fears, set forth in words of wisdom, so you don't surmise,
Hagridden by Religion, that the earth, the sun, the skies,
The sea, the stars, the moon, are bodies holy and sublime,
And made of godly stuff that must endure throughout all time,
And so you don't believe it should be treated as a crime
As heinous as the ones the Giants were guilty of, to bring
The ramparts of the world tumbling down with reasoning;
Or to wish to snuff the sun out, brightly shining in the sky,
By smirching such immortal things with rumours they will die.
These objects are so far removed from divinity, so at odds,
And are so little deserving to be numbered with the gods,
That, rather, they supply the paradigms for us instead
Of what lacks quickening movements and sensations and is dead.[14]

As Pramit Chaudhuri explains, in Lucretius, 'Gigantomachy becomes a symbol of independence, rationalism, and resistance to the superstitions and prejudices of the ignorant ... [Materialists are] capable of breaking into small pieces not only others' arguments but also the very fabric of the world – for what could better describe their theory of atoms?'[15] For Lucretius, it is no crime to bring the 'ramparts of the world tumbling down with reasoning'. Rather, this is human beings using their intelligence to conquer mysteries and assert their dominion over the world. So Plato's analogy could now be read in two ways. The idealists could be seen as upholding a conception of reality that transcends the physical world, rather as the gods upheld their order against the primal and chaotic forces of nature. Or the materialists could be seen as recognising only physical nature and seeking to tame it, in order to replace chaos with human order.

An ostentatious method of displaying dominance is to train your vassal to perform tricks for you. We do this with non-human animals: dogs beg, dolphins balance balls, chimps mimic people. We thereby make our supremacy among animals obvious. To assert our dominance over not just the animal kingdom, but the physical world, we use technology. In bending the world to our will, we make it serve us, and may thereby seek to improve the human lot. As the author of the influential *Pseudo-Aristotelian Mechanics* wrote, probably in the fourth century BC:

> One marvels at things that happen according to nature, to the extent the cause is unknown, and at things happening contrary to nature, done through art for the advantage of humanity. Nature, so far as our benefit is concerned, often works just the opposite to it. For nature always has the same bent, simple, while use gets complex. So whenever it is necessary to do something counter to nature, it presents perplexity on account of the difficulty, and art [*technē*] is required.[16]

To make things run counter to nature requires knowledge of nature, rather as an animal trainer must know his or her animals, and this knowledge was to be cultivated for the 'advantage of humanity'.

Plato's attitude to *technē* was mixed. He thought its best products were those which 'lend their aid to nature', but was opposed to any attempt to subvert the gods-given order.[17] These reservations might have been overshadowed when Aristotle's ordered and compartmentalising empirical research led to the first flowering of Greek science among his successors – such as Strato, whose materialist 'natural philosophy' incorporated rudimentary experimental methodology.[18] And yet, despite all the great achievements of subsequent Greek and Roman science, it eventually faltered. If it had steadily moved on to the likes of nuclear technology, we might not have made it past the Middle Ages, but it did not. Rather, as Benjamin Farrington puts it:

> there was no great forward drive, no general application of science to life. Science had ceased to be, or had failed to become, a real force in the life of society. Instead there had arisen a conception of science as a cycle of liberal studies for a privileged minority. Science had become a relaxation, an adornment, a subject of contemplation. It had ceased to be a means of transforming the conditions of life. Even such established arts as were adapted to keeping society in repair – professions like those of the architect and the medical doctor – were on the edge of respectability. They approached it only to the extent to which the practitioner could be regarded as the possessor of purely theoretical knowledge by which he directed the labour of others.[19]

So, the gods remained firmly in charge in ancient times, with their preference for theoretical knowledge over practical applications informing prevailing attitudes. Farrington illustrates ancient science's 'retreat from its function as man's weapon in the fight against nature', with some quaint examples of science being used to perform religious 'miracles'. For instance, worshippers who burnt their offerings on a specially designed alter could witness an icon of their deity burst forward to salute them – through the application of some of Strato's principles of pneumatics – while newlyweds could marvel at an iron Mars and loadstone Venus magically coming together in passionate embrace. The magic of science remained a novelty in the gods' world.[20]

The reason things did not stay like that is because we learnt to harness the 'magic of reality', as Dawkins aptly puts it in the title of his book for children which I mentioned earlier. The word 'magic' originates from the arts of the Zoroastrian *magi* (astrology, healing, etc.), which were held in suspicion by the Greeks. The Christians transferred the sinister connotations to Greek and Roman *technē*, on the grounds that it invoked pagan gods, who were really demons.[21] But by the time of medieval Europe, attitudes were softening, with a distinction now being drawn between 'natural' and 'demonic' magic. As Richard Kieckhefer explains:

> Natural magic was not distinct from science, but rather a branch of science. It was the science that dealt with 'occult virtues' (or hidden powers) within nature. Demonic magic was not distinct from religion, but rather a perversion of religion. It was religion that turned away from God and towards demons for their help in human affairs.[22]

Nevertheless, the term 'magic' was still often reserved for the demonic kind, even by philosophers who recognised the existence of natural 'occult virtues', such as Aquinas. The notion of natural magic gained more widespread acceptance among intellectuals in the fourteenth and fifteenth centuries, but even then, many continued to harbour a deep suspicion.[23]

The medieval distinction between occult and ordinary powers was sometimes, as Kieckhefer puts it, 'subjective', such that 'a power in nature that is little known and inspires awe is occult'. There was also a more objective sense, according to which occult powers were not internal to their bearers, but rather derived from the external source of emanations from stars and planets.[24] The objective sense turned out to be based on a false hypothesis. But if we understand natural magic as the science of occult powers in the subjective sense, it has never left us. Science today harnesses powers in nature that are 'little known', except to an expert community; and even there, the knowledge is fragmented between specialisms and specialists, and increasingly dependent upon machines. These powers inspire 'awe' and facilitate technologies which structure our lives. Thus, we push a button to store our digital photographs in a 'cloud', without knowing what a 'cloud' is. Our attention is focused on our goals, with the technology that delivers, moulds and frequently creates these goals remaining invisible; conspicuous in its awesomeness on first arrival, but not for long.[25]

Now we reserve the word 'magic' only for the fantasy of powers which science *cannot* explain; a fantasy which is fading fast, since we now tend to think the wizard's spell must, within its fictional world, have some scientific explanation. This transformative process began in earnest in the seventeenth century, at the beginning of which Francis Bacon made his highly influential case that philosophy must abandon its traditional, contemplative role in favour of a hands-on, practical approach. This was, in effect, a successful manifesto for science to eclipse philosophy.

Bacon's concern was with steady, incremental progress, bringing results which would change the world for our benefit. His disdain for philosophy's failure to achieve this survives unchanged in the anti-philosophy agenda today. Thus he thought, 'the wisdom which we have drawn in particular from the Greeks seems to be a kind of childish stage of science, and to have the child's characteristic of being all too ready to talk, but too weak and immature to produce anything. It is fertile in controversies, and feeble in results.' The studies he proselytised for aimed, 'not to defeat an opponent in argument but to conquer nature by action; and not to have nice, plausible opinions about things but sure, demonstrable knowledge'. To those sympathetic to his aims, he made the following entreaty: 'let such men (if they please), as true sons of the sciences, join with me, so that we may pass the antechambers of nature which innumerable others have trod, and eventually open up access to the inner rooms.' On accepting his invitation, we would find, 'as if awakening from a deep sleep, what is the difference between the opinions and fictions of the mind and a true and practical philosophy, and just what it is to consult nature herself about nature.'[26]

Bacon wanted to 'revive and reintegrate the misapplied and abused name of Natural Magic, which in the true sense is but Natural Wisdom, or Natural Prudence', finding in the word an 'ancient and honourable meaning' as 'the science which applies the knowledge of hidden forms to the production of wonderful operations'.[27] The extent to which Bacon, and the scientific revolution in general, was influenced by the occultist magical tradition represented by the likes of Cornelius Agrippa and Paracelsus is a subject of some debate; Bacon singled them out for vehement criticism, but shared their passion for making knowledge practical, observation-based and results-orientated.[28] However, in the sense of 'magic' that Bacon wanted to revive, rooted in the subjective notion of 'occult virtues', the scientific revolution for which he provided the definitive intellectual justification was to spread magic throughout our world. It changed the world just as he wanted, and now changes it ceaselessly. If we so wished, we could now have awe-inspiring holograms of our deities appear on altars, but our most revered places of worship remain ancient and unmagical.

It would be a mistake to conclude that the titans have won. Not just a mistake, but the wrong way of looking at it, one which would allow the adversarial nature of the myth to dictate the options for resolution. Magic, in Bacon's 'ancient and honourable' sense, has given us longer and happier lives, and there are now far more of us than ever before enjoying what this life has to offer. But our societies are governed by moral codes, and abstract ideas of fairness and equality have determined how we organize our societies. The gods and titans have both had a decisive role to play in our prosperity. Action and reflection are both important. The original analogy concerned reflection, since it referred to the philosophical debate between idealists and materialists. And magic is not a good argument in that debate: reason working on nature is what makes it happen, but it is not a reason itself. The frustration you see in Bacon, anticipated by Lucretius, still felt by the scientists discussed in the introduction, is that philosophical debates do not result in direct action; that philosophy is not science. But the philosophical questions remain, action cannot resolve them, and we will always need both reflection and action: both a direction and a mode of transportation.

If we allow ourselves to think the titans have won, and thereby allow an increasingly un-reflected philosophy to influence us, we may find science and technology steering our lives in a manner which individuals, when they reflect upon it, find that they do not want. The reflection may itself come to seem in vain, as I think is already happening, because it seems unscientific, and because the direction of travel seems inevitable anyway. And yet if we follow the myth through, the titans win when the gods lose control and chaos returns. It would be better to bring gods and titans into harmony.

§3. The Problem of Ceaseless Technological Advance

Human beings recently acquired the technological power to make themselves extinct within foreseeable, and hence frightening, scenarios. We might have made ourselves extinct before, in principle, but with great difficulty, and not within any realistic scenario. Extinction through disease or other natural causes might have occurred when our population was radically lower, as it was for most of our history, and it still might. But it would not be our fault and technology would be our defence. Nuclear technology, however, has given a standard name to the humanly caused endgame, 'World War III', a name which reminds us of the exceptionally violent recent history of our species, which goes far beyond that of any of the other animals we know about.[29] And if the next world war ever comes about, we will also have biological weapons to worry about. So that is already two resources we must hope the power with its back up against the wall, or just its aging, selfish leader, would never think of using. We also have to hope that no accidents occur, and that non-state organizations never acquire these capabilities – 'never' meaning over the course of centuries, or millennia for that matter. Another kind of technologically induced endgame we have even more recently started to imagine is ecological collapse. Since this has more similarity to a natural disaster than war, technology might be our best defence here too.

Would independent observers of the human drama think it wise for us to continue gaining technological power indefinitely and with increasing rapidity? Would they advise us to go full steam ahead, in light of our history of violence, the fact that our species lives in various divided and competing power blocs, and our track-record of weaponising new scientific breakthroughs? How confident would these observers be that we will still be prospering at the end of the twenty-second century, as we press ahead to what seems to be our Star Trek fantasy of exploring the galaxy to befriend other civilizations? Personally, I think they would advise us to ease off a little, and, during a period of philosophical consolidation, try to come to some agreement about what we are trying to do. There is nothing we are supposed to be doing.

There is a 'Doomsday Clock' posted by the *Bulletin of the Atomic Scientists*, an organization set up in 1945 by the scientists who gave us the first nuclear weapons. When I last looked, we were supposedly 'two and a half minutes' away from oblivion. Scientists who develop devastating technology seem to have acquired the habit of engaging in moral reflection after the event, inspired by Alfred Nobel – the inventor of dynamite – with his Nobel Peace Prize. Archimedes refused to produce any blueprints for the military technology he designed, so the knowledge could not be passed on.[30] The scientists who developed the bomb showed no such concern, and staggering as it is to contemplate, they were not even sure what it would do when they first trialled it: they entertained serious concerns about it causing an uncontrollable reaction with globally catastrophic consequences – but went ahead anyway.[31] Edward Teller, the 'Father of the Hydrogen Bomb', was able to draw some wry amusement from having taken this risk on our behalf, as well as from his popular epithet.[32]

But despite the political problem of nuclear proliferation that has afflicted our world ever since, and which shows no sign of going away, today's scientists may, for all anyone knows, themselves included, be preparing the ground for many new risks. And as seems to be the case with every scientific competition to make a 'breakthrough', they are doing so as quickly as possible; quite literally racing each other. As Teller said, 'Today's science is tomorrow's technology.'[33] Commercially viable artificially intelligent robots are not ready yet, but a new arms race in weaponized AI has already begun between the USA, China and

Russia. Meanwhile, plans are afoot for a sequel to the Large Hadron Collider, four times longer than the 27 kilometre original, so that even more of the awesome forces residing in subatomic reality can be uncovered and placed in the hands of humans as the twenty-first century develops.

We have created an atmosphere in which a scientist would not have to be a bad person to see a clear risk of their work being misused and yet still press on towards fame and fortune. Like professional athletes, they will have spent most of their lives immersed in their field of endeavour, simply to get where they are in such a competitive field; and mathematical arguments 'lack moral purpose', as Aristotle observed.$_{34}$ The genesis of the next techno-existential threat is easy enough to imagine without any malicious intent being involved. It has now become a blandly respectable idea that 'almost every civilization that develops a certain level of technology causes its own extinction', as Nick Bostrom puts it.$_{35}$ This is supposed to explain why we have not found any aliens: the extra-terrestrial species that became clever enough to blow themselves up promptly did so, thereby preventing them from ever becoming clever enough to develop the space-travel technology required to contact us.

The problem of ceaseless technological advance is the problem of keeping it ceaseless and beneficial. We are locked in a perpetual arms race by technological advance. We always have been. To stay safe, we need *our* power bloc to be the one with the best tech. But once you reach a certain level of destructive capacity, it no longer matters if only you have the cutting-edge tech, since inferior-spec capacity will be enough to bring everyone's story to an end. That is why the problem can no longer be ignored. The older the technology gets, the easier it is to get hold of; one day nuclear weapons will be as antiquated as Victorian technology is now, then comparable to Elizabethan technology, then medieval technology – but they will be no less destructive. Historically-entrenched rivalries in the developed world might still flare up, whether through unexpected disruptions, such as through global warming, or through shifts in the balance of power; China is likely to displace the USA as the preeminent global superpower at some point during this century.$_{36}$ And as worldwide wealth increases, and developing countries update their military capabilities, many more historically entrenched rivalries may escalate into existential threats. Within a few years of India and Pakistan acceding to the nuclear club in 1998, the Kashmir dispute generated one in 2001–2, and both countries are currently increasing their nuclear capabilities.$_{37}$ That we have already reached the point where ceaseless technological advance has emerged as a problem does not mean it is too late. It means it is time to do something about it so it does not get worse.

You might think the problem will solve itself, since new technology will permanently remove all existential threat; or that the problem is insoluble and all we can do is hope for the best; or that the benefits of largely unrestrained technological advance will always outweigh the risks; or that the risks have been exaggerated and are in fact steadily decreasing; or that we need to completely change human life by abandoning modern technology to return to nature. Or you might simply not care, on the grounds that there is nothing that you, personally, can do about it. But if we are ever to do something about it, we will need to think about it. And it is here that philosophy can play a role. For to form any kind of view on how we ought to be developing our technology so as to keep the advance both ceaseless and beneficial, we first need to reflect on what we ought, rationally, to want for our future. Different ideas for how to direct the advance must be brought together in rational debate. The more people get involved, the more collective rationality and consensual status the process can acquire. Rather than anticipated technological developments continuing to seem like worrying asteroids on a collision course with our planet, they might eventually start to seem more like policies to be encouraged or discouraged through argument, as well as through voting and purchasing decisions. Rather than wondering how the advance will change things next, we might learn to see it as a continually ongoing collective moral dilemma, to which expertise and experience from all walks of life can make a contribution.

I think that materialism is the main *philosophical* obstacle we currently face to learning to see technological advance as a collective decision, rather than as the result of political and free-market forces working with whatever science happens to turn up next. From the perspective of materialism, it becomes hard to see how a problem could possibly arise with science, when science is our method for finding out how the world is. From the materialist perspective, science is a matter of human beings uncovering as many truths about reality as they possibly can, thereby providing as much material as possible for technologists and politicians to put to use in solving our problems. Knowledge is power and you need as much of it as possible to solve problems. As such, materialism encourages the problem with ceaseless technological advance. And it discourages us from thinking about it, by rendering dubious any philosophical perspective from which we might seek to stand outside of the scientific description of reality to reflect upon it and ask how we should be developing it.

If knowledge is power, then it pays to know about materialism, and that is what the next two chapters will be concerned with. We may as well credit Democritus as its originator.[38] He was one of those audacious pioneers who aimed to work out, through abstract reasoning, the ultimate nature of reality. He reasoned that if you break things down to their smallest possible parts, then that is what everything is made of: 'atoms', he called them. Since that is what everything is made of, that is what reality ultimately is. This monumentally influential thought was titan-inspired, in that it offered us the prospect of technological power. What Democritus said made sense and still does, but it was the power we stood to gain by conceiving of reality atomistically which made his metaphysic so enduringly attractive.[39]

The gods' response to Democritus was to say: 'this, which you are breaking down into its smallest parts, this is consciousness – the immediate presence itself.' For the gods, the presence you are aware of right now is the ultimate reality – not the potential within it for gaining the power to satisfy our desires. Democritus's materialist philosophy had few followers until the seventeenth century, when its fortunes changed in the wake of the scientific revolution, then much more dramatically in the twentieth century, when the influence of religion over intellectual life declined and technological advance changed gear. Had materialism failed to catch on, science might have developed within a rationally restraining philosophical context. We might have seen that Democritus's insight was simply that reality can be treated in a broadly atomistic fashion, for the purposes of prediction and control. The notion of science as an inquiry seeking total truth, disinterested in any particular regard, might have remained alien to us. We might have seen that our ability to predict and control will not tell us how best to use that ability, and that the more our power increases, the more we need to reflect on our desires in the hope of regulating them with collective rationality. If Democritus's insight had been considered purely scientific, and not philosophical, then gods and titans might have learned to live in harmony by now. They still could.

2

The Materialist Philosophy

§1. The Metaphysical Issue

Idealists and materialists disagree about the fundamental, or ultimate, nature of reality. The words 'fundamental' and 'ultimate' mean the same in this context, but since 'fundamental' is also used in physics, I shall use 'ultimate' so that the issues cannot be confused, as they regularly are.[1] Materialism claims that the fundamental/ultimate nature of reality is discovered by fundamental physical science, and so by opting for 'ultimate' rather than 'fundamental', we remove any scope for forgetting that this is a philosophical claim, not a scientific one.

Now to claim that something is the ultimate reality is to claim that so long as it exists, then so will everything else we know about. This is not to claim that 'everything else we know about' does not belong to the ultimate reality, only that the conceptions we have of these things are not of the ultimate reality *as* the ultimate reality, but rather as something else: as something non-ultimate. Thus, experience and subatomic particles are candidates for the ultimate nature of reality, but dirty old boots are not. Nobody would suppose that the existence of all the dirty old boots of this world automatically brings in its wake the rest of the reality we know. Nevertheless, dirty old boots must belong to ultimate reality, since everything real must. It is just that nobody thinks we are focusing our minds on this nature as directly as we can when we think about dirty old boots.

Idealists think our inner mental lives provide the best model for trying to conceive the ultimate nature of reality. Materialists think scientific theories of matter provide accurate conceptions of the ultimate nature of reality. Idealists have a philosophy of ideas and materialists have a philosophy of matter. Thus, the contemporary materialist thinks that so long as the particles and forces described by our best physical science exist, there will be people, trees, rocks, and everything else we ordinarily think of as real. This is because reality is built up from, or constituted by, the basic physical matter. The contemporary idealist, on the other hand, thinks that given the existence of conscious experience, our own and that of other animals, there will be people, trees, rocks, and everything else we ordinarily think of as real. This is not because these things are *made* from experience – experience is not a kind of stuff – but rather because they are *implicated* by experience. We make sense of experience *as* experience of ordinary things like people, trees and rocks.

Everyone knows that experience and the physical world exist. The categories 'mental' and 'physical' provide our broadest and most secure conceptions of reality, so it is unsurprising that metaphysics would employ them in its dispute over the ultimate nature of reality. Although dualism tries to use both, the idealist and materialist keep it simple by just opting for one a piece, and it is their debate, the contemporary incarnation of Plato's 'interminable battle', that provides our interest here. If the idealist is right, experiential existence is ultimate and physical existence is not, and if materialism is right, physical existence is ultimate and experiential existence is not. For this debate to be rationally pursued, then, each side needs to account for the other's choice of ultimate reality as something which their own choice shows to be non-ultimate. The idealist needs to show that physical existence can be accounted for in terms of ultimate experiential existence, and the materialist needs to show that experiential existence can be accounted for in terms of ultimate physical existence. Then we can evaluate how well they both did and decide who we think is right.

These are the basic contours of the debate, as ought to be considered uncontroversial. And once we are clear about these contours, a disparity immediately emerges in the argumentative tasks facing the idealist and materialist. The materialist has a mountain to climb but the idealist does not. This is because in the very act of stating the idealist position, it becomes clear how they can approach their task. Their claim is that our experiences implicate the physical world, and since to talk about 'implication' is to characterize our minds, the idealist's candidate for ultimate reality provides the resources needed to fulfil their argumentative task. The details may be difficult, but we are just talking about the basic argumentative situation. In the case of the materialist, however, the idea of building up reality from its basic physical parts sounds fine for physical objects, but completely inappropriate for experience. Their candidate for ultimate reality is clearly very promising for dealing with things like boots, but barring some incredible conceptual innovation, not for experiences.

As we shall see in the next chapter, this conceptual innovation never transpired, so materialism naturally gravitated to the claim that experience does not exist. As any unbiased observer ought to conclude, then, they lost the debate. And yet materialism is the establishment position of today's academic philosophy and the metaphysical common sense of the secular modern world. From this position of dominance, materialists do not like to be reminded of the debate they belong to. Instead they like to think of their metaphysic as a scientifically established bastion against superstitious beliefs in ghosts, ESP and the like. But science does not favour materialism over idealism – both are fully compatible with everything science tells us about the world, and any philosopher who would dispute a scientific claim, as opposed to a philosophical interpretation of that claim, is a fool. Science might have discovered new evidence to inspire the conceptual innovation needed by materialism, but despite major scientific advances in correlating brain activity with experiences, the inspiration never came. Idealists would expect experiences to be correlated with the physical world they implicate, and there is no reason to think details of the particular correlations we have discovered, or will discover, have any bearing on the issue of how mental-physical correlation is generally to be understood: whether as a result of the physical being ultimately mental or vice versa. Individual scientists who write popular science books may often be materialists these days, as many people are, but that is just their philosophical conviction. Back in the 1920s and 1930s, the eminent physicist Sir Arthur Eddington was writing popular science books which incorporated his philosophical conviction in idealism.[2]

§2. Cold War Metaphysics

A striking fact about the history of philosophy is how sparsely populated it is by materialists. In terms of the standard big names, materialism has nothing like a Plato, Descartes, Kant or Wittgenstein to its credit. It does have Democritus, Epicurus, Hobbes and Marx; the ancients who gave us the idea and two highly influential political philosophers.[3] This is because materialism and atheism have been closely linked since ancient times, with interest in the former explicitly tied to the latter until very recently. Marx is famous for his antipathy to religion and Hobbes was regularly accused of 'atheism' in his day.[4] The association begins most clearly with Epicurus, who denied there is an afterlife and argued that the gods are indifferent to human affairs. Then, primarily through the influence of Lucretius, materialism became embroiled with the twin political themes of wresting authority from religion, and empowering ourselves with science and technology.

As Mary Midgley has emphasised, materialism began to take hold of the modern imagination by means of Lucretius's *poetry*, rediscovered in the Renaissance. In that poetic vision, materialism was presented as a 'moral crusade' which promised 'the only way to free mankind from a crushing load of superstition by showing that natural causation was independent of the gods'. What swayed Lucretius's new readers was his vivid portrayal of the terrors of the superstitious mind hoping in vain to appease the gods with rituals that, as materialists alone realised, were completely ineffectual on the natural order of things. Materialism promised peace of mind. Then after the 'wars and persecutions that disgraced the name of religion in the sixteenth and seventeenth centuries', Lucretius's enduringly resonant exclamation – 'how many crimes has religion made people commit' – struck a chord with Enlightenment philosophers. Materialism was 'once more seen as having a profound moral significance'.[5]

Marx argued that it was in the hands of the *philosophes* of the French Enlightenment that materialism first started to realise its revolutionary *raison d'être*. Marxists today still see materialism as a politically charged vision. Terry Eagleton, for instance, thinks it promotes solidarity with nature and with one another, as we feel part of a reality united by materiality; he also accepts that it may encourage us to think of nature and other people as objective resources to be exploited ('materialistic' behaviour, in the popular sense), but takes this to be a capitalist abuse of the doctrine.[6] Marx certainly thought materialism was essentially political, which is a thought that never seems to cross the minds of contemporary analytic philosophers. The materialism advocated by today's analytic philosophy, however, clearly corresponds with a stage in Marx's history: the one he associates with Hobbes, in which materialism aligned itself with modern science. The key innovation at this stage was that materialism now adapted itself to the theoretical world of Galilean science: a world that can only be understood mathematically. Galileo's physical world had no colour and subsequent developments in physics purged it of anything solid.[7]

This austere vision was necessary, according to Marx, in order to beat the priests at their own game. As he puts it:

> Hobbes was the one who systematized Bacon's materialism. Sensuousness lost its bloom and became the abstract sensuousness of the geometrician. Physical motion was sacrificed to the mechanical or mathematical, geometry was proclaimed the principal science. Materialism became hostile to humanity. In order to overcome the anti-human incorporeal spirit in its own field, materialism itself was obliged to mortify its flesh and become an ascetic.[8]

Marx thinks that the view that we have non-physical souls setting us apart from the physical world was, just like the ascetic form of materialism designed to combat it, a distancing, anti-human vision. Unlike materialism, however, non-materialistic metaphysics support ideologies designed to justify inequalities in the distribution of material goods and services. So, to beat them, materialism focused exclusively on the pure, theoretical description of the world provided by empirical science – to show that it had a better and more rational description to offer than that of the priests.

It is this kind of materialism which holds sway in contemporary analytic philosophy, as well as over contemporary secular common sense: the kind that tells us that if we want to know the ultimate nature of reality, then we must look to science. Ultimate reality is whatever contemporary physical science says it is. Neither advanced mathematics nor access to cutting-edge laboratories are required to be a materialist. There is no incompatibility between regarding materialism as common sense, and not being able to distinguish your bosons from your fermions. All you need is trust. Materialism could hardly be any more different from science.

Now it has occurred to many philosophers that contemporary materialism's essential deference to university science departments leaves it vacuous, because today's physics simply does not have a unified, final story to tell, and a future story can hardly provide our metaphysic if we do not know what it will be. Such considerations (standardly labelled 'Hempel's Dilemma') have led philosopher Alyssa Ney to argue that today's materialism is best construed as an, 'oath one takes to formulate one's ontology solely according to the current posits of physics', such as, 'I hereby swear to go in my ontology everywhere and only where physics leads me'.[9] If this is along the right lines, then Marx's view that materialism is essentially political looks more plausible than ever. It would seem that the trust people once placed in religious institutions was simply replaced with a new trust in scientific institutions, as part of an ancient moral crusade against organized religion.

That materialism is apolitical, however, has been taken for granted in analytic philosophy ever since its renaissance among British, American and Australian philosophers in the 1950s. This impression managed to emerge despite the political significance materialism had been accorded for the previous two thousand years, and which it was still being accorded in the Communist Bloc at the time, where nearly a third of the world's population resided. The last time materialism had enjoyed a significant and sustained measure of popularity outside of Russia was among the German materialists of the latter half of nineteenth century, who were fully upfront about their political motivation of wresting authority from the hands of religion to place it into the hands of science. Whether these now largely forgotten figures were philosophers or opponents of philosophy was a point of contention among them.[10]

Now although materialism clearly does make a metaphysical claim about reality which can be considered independently of its moral and political associations, the strength and persistence of these historical associations naturally raises the suspicion that the reason for its current dominance might not be of the right kind. The traditional appeal of materialism has been to free us of superstition and break the authority of religious institutions so that science can bring us happiness. As such, there are reasons to be a materialist which have nothing to do with metaphysical argument. If the reasons people became attracted to materialism were not metaphysical, then a very poor metaphysic might have slipped through to dominate our horizons. It is not as though the political and moral attractions have gone away, after all. Some 80 per cent of the world's population are religious and powerful cultural institutions reflect this fact.[11] Some of the most prominent contemporary materialists, such as Daniel Dennett and Richard Dawkins, show great dedication to undermining religion.[12]

Materialism has been around since ancient times without ever being the mainstream view it became in the twentieth century. And however favourably you might regard the pioneering philosophers who rehabilitated it in the journals – to be discussed in the next chapter – they made no spectacularly original theoretical moves. Materialist philosophy changed little, but the world in which people were philosophizing was changing a lot. Turbo-charged by the Second World War, technology was transforming life as never before, just at the time when materialism was gaining the dominance it has retained ever since. Technology was science making life better, so materialism seemed to be delivering on its traditional promise. Science had uncovered incredible secrets about the physical universe and religion was losing its sway over intellectual life. To young atheist philosophers, fascinated by and admiring of mathematics and science, religious idealist metaphysics was a living memory: the age of British Idealism continued well into the twentieth century, before positivism took the mainstream as a transition to materialism. There was the public profile of philosophy to consider. This new generation, which could see what was wrong with positivism but had still absorbed its message, did not want philosophy to be associated with religion, but rather science. Materialism was the pro-science and anti-religion philosophy. It was there on the shelf.

Witness Bertrand Russell, writing in 1925, some 30 years before materialism made its ascendency:

> At the present day, the chief protagonists of materialism are certain men of science in America and certain politicians in Russia, because it is in these two countries that traditional theology is still powerful.[13]

The perspective I am suggesting on the rise of materialism in the twentieth century is further encouraged when we note that just as few major philosophers advocated materialism before it became secular common sense, few did afterwards either. 'We are all physicalists now', David Papineau has said; although bolder than usual, this is typical of the kind of recruiting message today's materialists like to convey.[14] And yet the list of twentieth century analytic philosophers who either rejected or expressed serious doubts about materialism includes Russell, Carnap, P.F. Strawson, Putnam, Searle, Nagel, Kripke, McDowell and Parfit.[15] Not all philosophers deal with metaphysics, of course. So, who were the materialists of equivalent status? I think the only clear-cut cases are Quine and Dennett.[16] And you would be even harder pressed to identify major twentieth century materialist philosophers from outside of the analytic tradition. The only major continental philosopher to endorse a kind of metaphysical materialism was Deleuze, and materialism has only a minor track-record within non-Western philosophies, which was not influentially developed in the last century.[17]

§3. Disenfranchisement

Materialism sounds very plausible to people these days. Here is a way to make materialism sound plausible – a pretty standard one, I think:

> There is a physical world out there which exists independently of us. It obviously does, because things don't disappear when you're not experiencing them – they have a life of their own, as you notice when you accidentally bump into something. Science provides an incredibly accurate description of that independent world, and by breaking it down into its subatomic components, it has managed to get the whole thing within its description. Some philosophers and religious people think that in addition to all the physical stuff science describes, there are non-physical minds, souls, spirits, gods, etc. But this is obviously old-fashioned, superstitious nonsense: there aren't any spooky, immaterial things floating around in the physical world. People only ever thought that way because, being afraid of death, they wanted to have eternal souls and because they didn't know that our brains make our experiences. If experiences weren't physical, paracetamol wouldn't work!

And here is a way of making idealism sound plausible:

> Each of us lives through a stream of conscious experience, which is intermittently interrupted by sleep, and then, eventually, permanently ended by death. In a sense, that 'stream' is what our lives consist in; and the same must be true of other animals. Everything we know or care about either enters into our experiences, or else we believe in it in order to make sense of what does. Most of our experiences appear to be of an independent, objective world, and so in order to make sense of them, and tie them all together, we rely on this idea; we live by it and always have. By building on it with science, we have constructed a theory of the physical universe we experience. But we cannot have thereby produced the blueprint from which the ultimate reality could be constructed, because it only makes sense of what we experience, not experience itself; that is all it was ever *meant* to do. Experience is all we can be sure has independent existence,

because however we choose to theorise about it, it is still unavoidably there, at every waking moment of our lives and many sleeping ones too.

The key difference, of course, is that the first invites us to think about reality objectively and the second subjectively. Objective and subjective thinking are both natural ways of thinking, but by trying to extend them to achieve complete coverage, idealists and materialists make them less natural. Objective thinking is natural: when we look at a tree in broad daylight, nothing could be more natural than to think only of the tree. It is within this natural attitude that materialist thinking is born – and quite unnaturally, never leaves. For the naturalness fades of its own accord when we can only barely discern the tree's gloomy presence in the dark. Experience itself then becomes more obtrusive – and unmissable when we attend to the visual experiences we have when we close our eyes. When we cannot think of what we are aware of as something physical, we think of it subjectively: as experience. This is natural too.

Now there is one key thing to notice about the dispute between materialism and idealism as regards their social implications, and hence how accepting one or the other might bear on the problem of ceaseless technological advance. It is that idealism thinks of ultimate reality as something *within* each of our experiential perspectives, while materialism does not. Idealism thinks of it as something always there to be thought about, whether or not we can get anywhere close to thinking accurately about it. Thinking about ultimate reality in terms of its subjective, experiential appearance might be misleading, for properties such as being 'subjective' and 'experiential' may not really characterize ultimate reality. Thus, the idealist may think subjective thinking about reality accords with its ultimate nature only to the extent that it focuses our attention on the right thing and gets us thinking about *that* (which is my view, see Chapter 4). Nevertheless, however sceptical the idealist is about subjective thinking, their view is still that ultimate reality is to be found within each of our individual, experiential centres: it is what we call 'experience itself', rather than whatever it is we might be experiencing. Nothing could be more familiar than experience itself, so ultimate reality, on this model, is available for everyone to reflect on. Idealist metaphysics is democratic.

For the materialist, on the other hand, our experiences must be objective to exist, and given how strongly experiences and brain states are correlated, the materialist will typically hold that they are brain states, or somehow a product of objective brain activity. However, if the brain states we occupy are indeed there to be focused on from within our experiential perspectives, it does not seem that way. If we focus our attention on an experience, it will seem that we have *not* focused on a brain state. It will either seem that we have focused on something radically unlike a brain state, namely an experience, or that we have focused on some physical thing which the experience informs us of – it depends on whether we think about it subjectively or objectively.

And here we begin to see the disenfranchising vision of the materialist, according to which if we want to think about ultimate reality, then we should actively refrain from thinking of it as something we are experientially presented with. This is because if materialism is true, then focusing on what we are experiencing will result in distancing ourselves from ultimate reality. Attending to what we experience can only encourage us to think of our current brain state in a manner which makes it seem nothing like a brain state, or of the tree we experience as something coloured and solid. And so, because the ultimate reality can only be conceived abstractly by means of experimentation and mathematical reasoning, these experiential perspectives are a hindrance. The proposal is not that we think about what is experientially presented as objective rather than subjective; then we will find a coloured tree, not a colourless brain state. Rather, the materialist proposal is that we abandon, or intellectually disavow, our experiential perspectives when trying to think about ultimate reality.

So while idealist metaphysical thinking remains rooted within our own experiential perspectives, the materialist holds that the very act of focusing on what is experientially present to us is an act which conceals true reality. We misconceive reality simply in virtue of thinking of it as something we can point to, or fix our attention on. To think about it we need the theories that have been produced by experts abstracting away from a vast array of experiences had in response to various objective conditions in the world. Our own individual experiences are irrelevant, because reality can only be thought about in the abstract and you can think abstractly about anything at all – so long as you know the appropriate theory.

An essential part of thinking of yourself as possessed of an experiential perspective on reality is thinking that you have that perspective *now*. Reality is experientially *present* to you, just as it was in the past and will be in the future. However, since the materialist must defer our thinking about time to contemporary physics, they must reject as illusory the sense we each have of being at a particular stage of our lives; or, at least, try to resist the pressure to do so. For reality is four-dimensional, according to our best physics – time is a dimension of what it is. As such, assuming that the human objects are a distinctive enough part of reality to be worth talking about in a metaphysical context, which some materialists doubt, they are not the objects we can draw pictures of and photograph, but rather radically unfamiliar objects with time built into their very natures. They have a start (birth – or perhaps, conception), an end (death – or perhaps, decomposition), and a middle (what we might naturally think of as: our lives). Thinking that you are in the present is what situates you within the world and makes it your field of action. But if this is all just an illusion, as some materialists say and others try hard not to, then life as you know it is an illusion.

So it is not just that focusing on your current experience requires you to grossly misconceive the current reality of your brain state, for the very idea of the 'current reality of your brain state' is illusory too; or at least, to repeat the qualification, there is great pressure for a materialist to say this – much materialist philosophy, perhaps most, consists in attempts to resist these pressures imposed by the metaphysic itself. In any case, if you want to think about reality, it cannot help to focus your attention on anything you are experientially aware of, if to do so is to make an artificial and misleading abstraction from a reality which stretches indifferently throughout the time-span of your life. Any attempt to preserve the idea that you are living that life subsequently becomes an exercise in dubiously motivated philosophical gymnastics. For if we are not living our lives in anything like the sense we have always imagined, then it is hard to see what motivation there could be for inventing a new and legitimate sense of 'living our lives'.

Metaphysical understanding is bound to relegate much of our ordinary understanding to non-ultimate status, by telling us we are thinking about things not as they ultimately are. That is the business it is in – if you focus on the ultimate, then that will be the consequence. But whereas idealism tells us how to think about the ultimate reality of the non-ultimate realities we weave our lives around, materialism puts even their non-ultimate status in question. Is there any justification to calling a four-dimensional reality of subatomic particles and forces 'a tree' or 'me looking at a tree?' Is there any justification to calling a four-dimensional reality of subatomic particles and forces 'a practical justification'? Once our experiential grip on life is abandoned, these become open questions, and any pressure we feel to answer them in the affirmative becomes a cause for suspicion. This actively discourages autonomous philosophical thought, and encourages a conception of philosophy as an exclusively professional pursuit: another kind of science, and, as outsiders might very naturally think, a completely pointless one.

§4. Scientism

Materialism provides the intellectual bedrock of scientism. Scientism tells us to let scientists do the thinking for us, and this attitude is creeping into the law, art, economics and the humanities generally. Raymond Tallis nicely sums up the currently dominant agenda as follows:

> If we are our brains, and if human societies are the summed activity of brains interacting with one another and with the material world, it follows

that everything we do individually and collectively can be understood in terms of neural activity. The findings of neuroscientists, supplemented – since the brain is an evolved organ designed to help us face the exigencies of life in the jungle or the savannah – by evolutionary thinking, will reveal the true nature of our behaviour and the institutions that regulate it, our customs and practices, norms and laws, arts and sciences. If you want to understand people, look at their brains. The writing is on the wall and the script is pixels on a brain scan. Roll over, social sciences and humanities, allow yourselves to be incorporated into a vastly extended neuroscience and discover your true nature as animalities.[18]

Even if making connections from what is going on in our brains to our lived realities is a spurious business – as Tallis and other anti-materialists argue powerfully that it is – once the materialist picture is presupposed, arguments to this effect must always be on the defensive and under suspicion of being merely reactionary. Reason itself is on the defensive, except when it takes its beginning from an interpretation of scientific evidence. Even if an argument makes complete sense to us, if it contradicts something far less compelling which a team of scientists say on the basis of their research, there is cause for doubt. Making sense has already started to seem like no competitor to scientific truth. We have learnt to marvel at the lack of ordinary sense which quantum mechanics makes, primarily through the Schrödinger's Cat example, without doubting for a second that it is true; some 30 per cent of the United States' GDP relies upon that science, after all.[19] It is certainly true that it works. If we come to think that all truths are scientific, then ordinary, active reasoning, which is guided by what makes sense to individual people, is liable to wither or mutate along post-truth lines, which I think is what we are already seeing (see Chapter 8). The 'simplest and most straightforward sign of scientism', as Susan Haack puts it, is 'being too ready to accept anything and everything bearing the label "science," or "scientific," and to believe any and every claim made by scientists of the day'; it is the 'epistemological vice of credulity'.[20] This credulity is crushing non-expert reasoning while fragmenting the rest, since scientific reason is fragmented among a vast array of experts within narrow fields; and when the credulity collapses, an extremely damaging distrust of science results. Science itself has not caused this problem, but rather the materialist interpretation of science and the lack of philosophical reflection it engenders.

There are many ways to challenge scientism. History can put the phenomenon into a wider perspective and so can the law.[21] But materialism is philosophy's business. And it is a crucial one, because while we might recoil from the thought of past scientific justifications of social policy, most notably eugenics, and while we may seek to moderate technological advances with legislation, the fact remains that while we believe ourselves to live in a world which is described without remainder by physics, the thought of using scientific knowledge to steer all aspects of our lives will never go away. That is the exactly the kind of background and typically un-reflected belief by which philosophy exerts its influence. While materialism remains in place as the common-sense philosophy of the modern secular world, however un-reflected, the reason of those who *claim* to have science on their side will have the most weight. As far as philosophy is concerned, materialism is the main obstacle to rational reflection on the development of our technological capacities. We cannot question the scientific understanding of the world that facilitates technological advance if that is the only understanding we are allowed to use.

§5. Weariness

It was from within our ordinary lived realities of day-to-day life that the ideas of freedom, consciousness, the self, morality, love, and other matters of human and philosophical interest arose. Materialism inspires the thought that they all arose from an illusory conception of reality. Given the view that reality is the objective one described by mathematical physics, the suspicion was bound to arise that we are not really free, conscious, or anything physically significant enough to call a 'self'. Some materialists come to these conclusions as if they are revelations, while others fight valiantly against them, but the agenda set by materialism is clear enough. By accepting this metaphysic you immediately put our ordinary way of looking at things into question, so the quest to preserve that picture undertaken by moderate materialists – who I think have always been the majority – strikes me as suffering from a very serious problem of motivation. An insoluble one, I think, because it shows a trust in the ordinary picture which is exactly what the metaphysic should have undermined. I think that the more hard-core and uncompromising the materialist, the more likely they are to have seen the real implications of their position.

It is because philosophy begins and ends in concepts rooted in our lives that the endpoint of the materialist endeavour cannot be a final philosophy enjoying the kind of solid consensus sometimes achieved in science. Such an ending would escape the divergence of rational opinion which is characteristic of the philosophical tradition and anathema to scientism – but since materialism is a philosophy, things could not end that way. Rather, if left unchecked, the endpoint of materialism could only be reached when we no longer reasoned about the metaphysical significance of scientific descriptions of the world, and no longer gave intellectual credence to any other kind of description. The endpoint of materialism is to forget itself.

There can be no long-term future for materialism as an active philosophy, rather than an embedded belief-system. Materialist philosophers might envisage their future as one of reasoning about the latest scientific results, as I think most do. But there is no reason anyone should listen to their take on the truth science uncovers when they are the ones telling us that only scientific methods are to be trusted. If we need interpretations, and they cannot be provided by yet more science, then why not leave this task to the scientists, who are bound to know more about the matter? What qualification could someone who identifies with a tradition of pre-scientific ignorance possibly have to pontificate on such matters? Unless materialists start producing their own scientific results, and hence become scientists, then they are doomed, by their own criteria, to eventually be dismissed as charlatans. Dennett, who is a cognitive scientist *and* philosopher, has come to disparage philosophy with 'no aspiration to make novel predictions'; in other words, philosophy that is philosophy rather than science.[22]

Few materialists seem to be aware of this direction of travel, but Alex Rosenberg is. He ended a piece in *New York Times*'s philosophy column in the following, striking fashion:

> What naturalists really fear is not becoming dogmatic or giving up the scientific spirit. It's the threat that the science will end up showing that much of what we cherish as meaningful in human life is illusory.[23]

'Philosophical naturalism' is a nebulous label. It often just means materialism, but a philosophical naturalist could simply be a philosopher who pays close attention to the results of the natural and social sciences, and who rejects any kind of supernatural explanation; although if the latter is the concern – spooks and telekinesis – then it seems to me that they are pushing against an open door. Such a philosopher might reject materialism, and some do.[24] But in the sense Rosenberg intends, philosophical naturalism is the view that only science delivers genuine knowledge of the world – all other supposed knowledge must be hedged, amended, or dismissed. Naturalism in this sense goes hand in hand with materialism: they are the epistemology and metaphysics of deference to science.

Rosenberg made this statement in response to the charge, effectively made by Timothy Williamson, that naturalism is dogmatic.[25] Naturalism is anything but dogmatic, Rosenberg says, since it bravely faces up to the possibility that, 'much of what we cherish as meaningful in human life is illusory'. Dennett has issued similar warnings: 'watch out – I'm coming after some of your most deeply cherished intuitions'.[26] There is pride behind this kind of statement: pride in being engaged in the disinterested search for truth. So, Rosenberg makes his final rebuke to Williamson by admitting that he is not *personally* disinterested: he fears the outcome. His fears were soon to be realised, however, since almost immediately after saying this he published, *The Atheist's Guide to*

Reality, which argues that freedom, consciousness, the self, morality, and anything else you might care to mention – anything apart from fermions and bosons – are illusory. Matter is composed from fermions, fields of force are composed from bosons, and physics is '*the whole truth about reality*'.[27] Reviewers have complained that the title of Rosenberg's book is misleading, since it contains little about atheism *per se*, but I tend to disagree.[28]

When the influence of religion on intellectual life started to decline, materialism offered philosophers something new to trust: science. A deep and attractive affinity between religion and science was sensed. Just as religion had seemed to offer closure on issues for which gods, or the will of the gods, could be invoked, now science seemed to offer closure through its ability to solve its problems and move on. But theological interpretation never ends, and the progress of science need not be thought of in this materialist manner, but rather as a matter of building up a stockpile of data so that future interpretations have more hard evidence to go on. Nevertheless, now that religion seemed less attractive to philosophers, the materialist merging of science and philosophy seemed to place the prospect of scientific closure within the reach of philosophers. Believing in nothing apart from fermions and bosons, as Rosenberg does, is what you might call the *ultimate* scientific closure for a philosopher.

It is not hard to see why this might seem attractive. Judged by the standards of closure, philosophy has had little or no success: because it still asks its ancient questions. But these are not the standards of philosophy. By its own standards, it has succeeded magnificently because it *kept the conversation open*. We never let it close down, and so have managed to maintain a rational conversation that stretches throughout our history; one anyone can join, at some level at least, without the need for specialist training or equipment. That conversation pervades our history: idealism, materialism, scepticism, stoicism, rationalism, empiricism, romanticism, utilitarianism, feminism, postmodernism – these are just some of the many currents of thought which were born of philosophy and which have had an incalculable influence on people's lives, for better or for worse. Philosophy is a conversation of intellectual freedom: one which answers to natural curiosity, relies on ordinary reasoning and sense-making, and concerns reality itself. But it is inconclusive, never-ending and wearying.

Materialism rejects philosophy's model of openness in favour of scientific closure. This may come to seem attractive as a result of frustration at the fact that philosophy's own kind of closure is usually just tentative personal intellectual satisfaction, and acceptance in the philosophical community that your view is worth discussing. Such frustration is with the nature of philosophy itself, its essential openness and historicity, and the fact that it typically concerns itself with matters about which decisive evidence is not to be had. Philosophy's own kind of closure was never a major concern to it, or should not have been – except as a spur to individual effort. But by focusing too much on it, the kind of closure science seemed to offer materialist philosophy became appealing. The root of this was weariness: a lack of will to keep the conversation going. Perhaps it was because of where the conversation had been going lately – the so-called 'death of God' brought out major anxieties in twentieth century continental philosophy, and perhaps materialism was analytic philosophy's more repressed answer. Or perhaps labouring away at philosophical questions suddenly seemed far too long-term and inconclusive in a world where science was achieving such rapid and definitive results. But whatever the reasons, materialism's attempt to merge scientific with philosophical closure ultimately results in anti-philosophy, of the kind expressed by the scientists I discussed in the introduction. Scientists will continue trying to solve scientific problems, and sometimes succeed, whether or not the closure they achieve is called 'philosophical' by materialist philosophers.

Rosenberg says, '*the science* [my emphasis] will end up showing that much of what we cherish as meaningful in human life is illusory'. Not so, and by his own admission too. For in the autobiographical preface to his book, he tells us that he gave up on a career in physics because it 'didn't scratch the itch of curiosity that was keeping [him] up at night', and so turned to philosophy instead. It then took him forty years to work out his answers.[29] And indeed, science will never tell us that freedom, consciousness, the self, morality, etc. are illusory. Materialism will. Science can tell us about the bodily behaviour, biology, chemistry, neurophysiology, etc., which transpires when we talk about such things, but it is materialist philosophy which inspires people to place this philosophical spin on the science. When they do this, it is because they think their materialism commits them to doing so, and I agree that it does. But it can also be a personal relief.

Rosenberg underlines this point in the conclusion to his book, where he says that if we find the disillusionment of materialism too hard to bear, then the answer is to, 'Take two of whatever neuropharmacology prescribes'.[30] The existential unease that attracted him to philosophy, then, is best solved by drugs designed to get the fermions and bosons back in order. Rather than go through all that he had to endure, however beneficial his efforts might prove to others, we could instead take the most direct comfort in the power of science which is currently available; as philosophical redaction surgery is not.

Analytic philosophy has a track-record of this kind of thing. Consider its great icon: Wittgenstein. Philosophical questions tormented him, making it natural for him to write that, 'The real discovery is the one which enables me to stop doing philosophy when I want to'.[31] Perhaps his personal life would have gone better if he had been able to follow Rosenberg's advice. But although he was no materialist and would no doubt have found the advice grotesque – he worked at philosophy right up until the end of his life – his captivating persona, if not his philosophy, helped set the scene for what we are experiencing now. The rise of science and decline of religion in the twentieth century made some philosophers weary. Materialism proposes the cure of falling uncritically into the arms of science. But a much better cure is to regain confidence in natural philosophical curiosity and the various traditions of reasoning it has inspired throughout our world.

3

When Philosophy Lost Its Mind

§1. Is Materialism Plausible?

Remember my attempt to make materialism look plausible in the last chapter? I am familiar with the materialist mind-set because when I started out as a philosopher I was a materialist. I hardly needed any persuading once I first learned what the metaphysical options were, but nevertheless these are the kind of considerations that used to maintain my conviction that materialism is obviously true:

> There is a physical world out there which exists independently of us. It obviously does, because things don't disappear when you're not experiencing them – they have a life of their own, as you notice when you accidentally bump into something. Science provides an incredibly accurate description of that independent world, and by breaking it down into its subatomic components, it has managed to get the whole thing within its description. Some philosophers and religious people think that in addition to all the physical stuff science describes, there are non-physical minds, souls, spirits, gods, etc. But this is obviously old-fashioned, superstitious nonsense: there aren't any spooky, immaterial things floating around in the physical world. People only ever thought that way because, being afraid of death, they wanted to have eternal souls and because they didn't know that our brains make our experiences. If experiences weren't physical, paracetamol wouldn't work!

This is just one big misunderstanding. You can pick holes in anything, if you are prepared to be pedantic, but there is nothing to be said for any of the above, as I shall now try to show.

We see the first move being made with the common-sense claim that there is a physical world existing independently of us. But it is equally a matter of common sense that there are experiences existing independently of us – most obviously those of other people, or, if the 'us' means human beings, then those of cats and dogs. Our experiences do not exist independently in the sense that we freely exercise control over them, but exactly the same could be said of our physical bodies, so if this disqualifies our experiences from independent existence, then our bodies do not belong to the independent physical world either.

To invite us to think that our conviction in the independent existence of the physical world has anything to do with materialism is to try to equate 'independent existence' with 'ultimate reality'. But we are not always doing metaphysics. And the plan backfires anyway, because we also think of experiences as having independent existence. All we are doing is registering the fact that both physical objects and experiences have their own self-sustaining existence, unlike a fictional character like Gandalf, or the things we see in dreams. By 'independent' we basically just mean 'real'. The debate between idealism and materialism concerns how this reality is best explained at the ultimate level.

The second move is to draw our attention to the success of science in producing a comprehensive and explanatorily powerful account of physical reality. But both idealists and materialists accept this. Materialism cannot claim the support of science, and certainly cannot claim any credit for its extraordinary explanatory successes, when it so often seeks to interpret those explanations as revelations of illusion. The materialist could try to turn this move into an argument by saying that given the methodological superiority of science as a means of investigating the nature of reality to any other methods we have at our disposal, then if anything deserves to be described as 'ultimate reality', the physical universe does. But that just ignores the sense of 'ultimate' which makes sense of the metaphysical debate. The materialist is now misconstruing this term as an accolade for the reality captured by the most detailed, explanatorily powerful theory we have. If that were the sense of 'ultimate' in dispute, there would never have been a debate in the first place. Even before we had modern science, it would have been obvious to anyone concerned that objective thinking is the detailed kind we rely upon for most of our everyday dealings. Plato would have agreed that the physical world is the ultimate reality, if that is what was meant.

Then we get to the third and most important move. Aware that the issue hinges on the nature of experience, the materialist tries to discredit our natural, subjective understanding of it. So, they invite us to look at the world objectively, and point out how bizarre it is to suppose that it might be populated by anything other than physical things. Non-physical minds look just like physical things as we are now invited to think about them, yet infinitely more dubious. We are invited to imagine them floating amorphously around the physical world, like ghosts.

This is hardly a fair characterization of the opposition, undertaken for the purposes of engaging in rational debate. If minds are *non-physical*, then it cannot be right to think of them as floating around in the physical world. If they are non-physical they are not part of the physical world, and hence neither a normal nor a dubiously wispy part of it. Any philosopher who ever seriously contended that minds or experiences were non-physical was not thinking about them objectively, but rather subjectively – in terms of how they appear to the subject who has them. This is particularly hard to miss in the case of Descartes' famously first-personal meditations.

To sum up, then, all we have seen is: (1) an invitation to think about the world objectively but not subjectively, despite the fact that we naturally think about it in both ways; (2) admiration for science combined with an attempt to convert the contested accolade of 'ultimate' into an expression of that admiration; (3) an apparently wilful misreading of the opposition. But intuitions and plausibility, important though they are, can only take you so far. In the last chapter I provided reason to suspect that the ascendency of materialism to its current position of dominance in the twentieth century might not have been rational. The only way we can be sure, however, is by looking at what the professionals said at the time. I shall begin with an historical overview in the next section, before moving onto the philosophical details.

§2. Two Paths to the Standard Picture

Analytic philosophy, like the aeroplane, was invented at the beginning of the twentieth century. It was a reaction to idealism, which during the nineteenth century was often heavily embroiled with religion; not always, however, as is most clearly testified to by the vehemently atheistic idealist, Schopenhauer. Analytic philosophy, with its focus on logic and language, combined with a piecemeal rather than systematic approach to philosophical problems, succeeded in making philosophy look more like an exact science than a theological disputation, which is something which evidently struck a chord at the time. The rise of materialism can be traced through two distinct paths stemming from this source, which came together in the 1950s.

The first came via 'ideal language philosophy', which sought closure on philosophical problems by removing the ambiguities in natural language that create the illusion of genuine problems: in the ideal language, there would be no philosophical problems. The ultimately anti-philosophical goal of this programme allowed it to organically merge with nineteenth century positivism to produce logical positivism. The central task of logical positivism was to relate everything that could significantly be said to sense-data reports – that is, reports of what we would ordinarily take to be experiences. The logical positivists wanted us to bracket this assumption, however, and instead just take it for granted that we receive a certain kind of evidence (sense data), the nature of which cannot be scientifically questioned and hence should not be questioned at all.

This was a tactical error for an anti-philosophical agenda. For although the idea of pure, unquestionable data, which originated in Ernst Mach's phenomenalism, is what motivated the new terminology of 'sense data', their metaphysical status always did look rather puzzling.[1] If you do not want people to wonder what is in the box, it is probably best not to make the box your central topic of conversation. So quite apart from the internal problems with logical positivism – the most notorious of which being that it had no significance according to its own criterion of significance, and that the project it called for turned out to be impossible to enact – it had the external problem of being an unsuitable medium for reflecting the creeping anti-philosophy of the age.

Quine was the greatest critic of logical positivism's internal problems, but his landmark 'Two Dogmas of Empiricism' article of 1951 laid the grounds for addressing the external problem too. As regards the internal problems, Quine attacked the logical positivists' project of finding analytic synonymies between sense-data reports and other statements, together with the general idea that all meaningful statements could be reduced to such reports. His conclusion was that our theories are only answerable to 'experience' as a whole – this was the word he used to replace 'sense data' as the ultimate grounding for our theories. According to Quine's favourite simile, we are like sailors repairing our ship at sea, with the 'ship' being the theory we must rely upon to make sense of experience and thereby avoid sinking.

As the 1950s progressed, however, it became clear that Quine's ship was science. Soon, his holistic talk of 'experience' was replaced with 'surface irritations', because science does not describe sense data or experience, but it does describe effects on our sensory organs. Irritations could be just as particular as sense data, and played a similar role in his thinking; but Quine retained the view that reports of them could not determine a unique theory. This residual philosophical flexibility, together with his relative disinterest in consciousness, kept Quine's views at a certain distance from contemporary materialism, as he continued along the lines of the old programme to devise an ideal language: by trying to get the language of the physical picture into good logical shape, so we could better see its true ontological commitments. By this point, philosophers could fly around the world to conferences, as Quine avidly did.

The other path came from the 'ordinary language philosophy' offshoot of the analytic movement, which sought to dispel the illusion that there are genuine philosophical problems by paying closer attention to how we ordinarily use language. The leading idea was that the philosophical tradition had created its own artificial language which generates artificial problems. Just as with ideal language philosophy, an ultimately anti-philosophical goal combined with apparently important work for young philosophers to do in the present.

Now earlier in the century, John Watson had revolutionized the fledging science of psychology by arguing that it should disregard introspective reports of our states of consciousness to focus exclusively on objective, observable behaviour. Gilbert Ryle, a leading figure of ordinary language philosophy, adapted this idea to argue that Cartesian dualism, which is commonly understood as holding that the mind and body are separate substances, resulted from a 'category mistake': that of supposing that to talk of a person's mind is to talk of something belonging to the category 'thing', when it is really just to talk about their behaviour.[2] This drew new attention to the metaphysical status of mind.

In the 1950s, a young, English psychologist named U.T. Place abandoned Rylean behaviourism because he did not think it could account for the life-changing potential of religious experience; it appears the gods were trying to bring things back into focus.[3] Conscious experiences could not simply be Rylean dispositions to behave, he thought, because there was a current reality to account for. As such, he produced a novel and highly influential version of the traditional materialist view that conscious states are brain states.[4] His colleague, J.J.C. Smart, was a great admirer of Quine and his materialism. And it was now that the two paths came together. At the end of the decade, Smart published a defence of Place's position which attracted immediate and extensive attention from philosophers determined to prove him wrong.

Smart was the key figure in establishing materialism's hegemony within analytic philosophy, so it seems rather ungrateful that his name is not revered by contemporary materialists; he is generally considered a rather minor figure these days. It was Smart who was responsible for what Daniel Stoljar has called the 'Standard Picture': a new mission statement for analytic philosophers.[5] The Standard Picture told philosophers that materialism was something we have 'overwhelming reason to believe', as Stoljar puts it, but which is 'prima facie in conflict with the central claims of human life'.[6] Analytic philosophy's new mission was to resolve this conflict. This mission had evolved from the explicitly anti-philosophical missions of both ideal language philosophy and ordinary language philosophy, and its aim was to avoid Rosenberg's conclusion that the conflict is actual, and hence that if materialism stays, the 'central claims of human life' must go.

Smart's mission held great appeal across the board. It provided some philosophers with exactly the kind of job they were looking for: one which gave them the feeling of lending a helping hand to science, and which made it plain for all to see that you could be a philosopher without having any patience with religion. Simply not believing is never enough for some people. Meanwhile it provided other philosophers with a good target. Both materialists and anti-materialists had their work cut out for them by Smart's Standard Picture, then. But if there never was any 'overwhelming reason' to believe materialism, then perhaps it is high time for analytic philosophers to cut their losses, forget the Standard Picture, and stand up against anti-philosophy.

In the next three sections, I shall provide three mutually reinforcing reasons why materialism should be rejected. Firstly, I will show that there was never any good reason to take it seriously in the first place ('The Lacuna'). Materialism was taken exceptionally seriously, however, and yet nobody ever managed to make good on it ('The Let-Down'). And when we follow through on where these efforts have led us, we find materialism evidentially undermining itself: if it were true, we could never have had reason to believe it in the first place ('The Loop'). This final, definitive problem was known to Democritus himself, who, after proposing his theory as a product of intellect, had his senses reply: 'Poor thought, do you take your warrants from us and then overthrow us? Our overthrow is your fall.'[7]

§3. The Lacuna

Quine is arguably the most significant non-political philosopher that materialism has to its credit and he was instrumental in bringing about its twentieth century renaissance. A prime and early example of Quine's materialism is 'The Scope and Language of Science', published in 1954. The opening sentence is, 'I am a physical object sitting in a physical world'. He goes on to describe how in response to irritations on the surfaces of this physical object, it emanates air-waves, i.e. he talks.[8] Quine then provides some justification:

> The quest for knowledge is properly an effort simply to broaden and deepen the knowledge which the man in the street already enjoys, in moderation, in relation to the commonplace things around him. To disavow this very core of commonsense, to require evidence for that which both the physicist and the man in the street accept as platitudinous, is no laudable perfectionism; it is a pompous confusion . . .[9]

Quine is thinking of his anti-foundationalist argument in 'Two Dogmas': that we cannot construct knowledge from a presuppositionless starting point, but must rather rely on the theories we already have (our 'ship').

Let us grant that our knowledge of anything is indeed 'theoretical', in the sense that we must understand it some way or another.[10] We cannot escape from theory, in that sense. Still, it is just as much part of common sense – just as platitudinous – that we have experiences as that there are physical objects. We naturally think both objectively and subjectively. We already have two kinds of 'theory', then, and we always did – hence the ancient standoff between materialism and idealism. Rejecting non-theoretical foundations to knowledge has no bearing on this issue.

Earlier, in 1948, Quine can be found talking about two 'competing conceptual schemes, a phenomenalist one and a physicalist one', and saying that, 'each deserves to be developed'. The physical scheme is simpler because 'Physical objects are postulated entities which round out, and simplify our account of the flux of experience', but the phenomenal is epistemologically 'fundamental'.[11] Later, in 'Two Dogmas', when he definitely turned against phenomenalism, he said that the 'myth of physical objects is epistemologically superior to most in that it has proved more efficacious than other myths as a device for working a manageable structure into the flux of experience' – but he still describes the conceptual scheme of science as only, 'a tool, ultimately, for predicting future experience in the light of past experience'.[12] An idealist might say exactly this. And yet by the time of the 1954 paper, the word 'experience' is notable for its absence. Now we have only surface irritations.[13] What changed?

The answer stems from Quine's heritage in the ideal language programme. The task which always drove him was to construct an idealized, fully systematic language to take us away from the vagaries of natural languages, and thereby enable us to see the ontological commitments of the simplest theory needed to explain everything in the world. Once he had decisively broken with phenomenalism and sense data, he pursued his systematising efforts within science, conceiving them as continuous with it. The simplicity of the hypotheses being systematized was key for Quine and counted as a 'kind of evidence', since 'scientists have indeed long tended to look upon the simpler of the two hypotheses as not merely the more likable, but the more likely'; 'likeable' was also a significant factor for Quine, however, since he famously had a 'taste for desert landscapes'.[14]

So, ontology, and hence the ultimate ontology of metaphysics, was to be decided by the most sparsely committed and systematic language, and Quine thought that once experience was stripped of any special epistemic role, there was no longer any motive for including it. This reasoning, which seems to have been decisive in determining his turn to materialism, was most clearly laid out in 'On Mental Entities' (1952):

> Epistemologists have wanted to posit a realm of sense data, situated just me-ward of the physical stimulus, for fear of circularity: to view the physical stimulation rather than the sense datum as the end point of scientific evidence would be to make physical science rest for its evidence on physical science. But if with Neurath [who originated the 'ship' simile]

we accept this circularity, simply recognising that the science of science is a science, then we dispose of the epistemological motive for assuming a realm of sense data.[15]

He goes on to say that, 'we decide what things there are, or what things to treat as there being, by considerations of simplicity of the overall system and its utility in connection with experience, so to speak.' However, this is only 'so to speak', as he immediately points out, because 'it is moot indeed whether the positing of objects of a mental kind is a help or a hindrance to science [...] moot or else it is clear that they are a hindrance'. And it was Quine's 'hypothesis, put forward in the spirit of natural science', that experiences and other mental states were in fact a hindrance.[16]

Thus, Quine thought it was only a scientific hypothesis that sense data existed; one made by scientists for the purposes of providing non-circular epistemological foundations for science. Consequently, as he seems to have reasoned, it follows that if that project was misconceived – as he had come to think it was – then we do not need sense data. This seems to be his main argument for materialism.

As regards the logical positivists, Quine's reading of the situation seems basically correct: some of them were scientists, and providing evidential foundations for science was a key role they had in mind for sense data. Another, however, was to eliminate metaphysical questions – which was not so much of a concern for Quine, given that he did not recognise any distinction between science and philosophy anyway; in this regard, you might say that he was positivism's greatest legacy.[17] However, in order to convert scepticism about sense data into an argument for materialism, Quine has to presuppose that sense data and individual experiences are the same; and here his analysis is incorrect. The positivists may well have had experiences in mind when they formed their conception of sense data – it is hard to see what else they could have been thinking of – but that does not mean their conception was accurate.

Experiences were not scientifically or even philosophically hypothesised to epistemically ground science. At most, a certain conception of them was hypothesised by the positivists. Experiences are as much a part of our everyday lives as physical objects. They are part of a primordial 'theory', if you want to think of it that way, but are they a 'hindrance to science'? The science of consciousness is currently doing very well in its project of finding out which experiences correlate with which brain states. Experiences do not seem to be a hindrance to science, then; almost all working scientists, I assume, take it for granted that their work revolves around experience. Experiences are certainly a hindrance to materialist philosophy, however, since their manifest existence is the main problem it faces. Quine can hardly argue for materialism, then, by treating them as if they were questionable scientific posits that have proven ineffective in the task they were designed for – since this is patently false. And he can hardly argue for materialism on the grounds that since experiences hinder materialism, we should deny their existence in order to pave the way for materialism. You cannot argue for atheism by saying we should deny the existence of God to pave the way to atheism.

So, is ontology to be determined by the sparsest systematizing language we can formulate, as Quine thought? Appealing to simplicity as a principle of evidence has been a controversial idea in philosophy ever since mediaeval times, and, *contra* Quine, it is controversial among scientists too.[18] It is not hard to see why, unless you are prepared to assume that reality was designed to fit the simplest human theories. But one thing is for sure: the systematizing language must include *everything*. If the desert has an oasis, then the metaphysician is obliged to mention it. Quine thought a materialist system could indeed include experience, and so, in the 1954 paper, he quickly dispatched this issue – the main one for any would-be materialist – with a 'facile physicalisation of states of mind'. He said that, 'An inspiration or a hallucination can, like the fit of ague, be identified with its host for the duration'; we simply identify the 'x' in question with 'the corresponding time-slice x' of its physical host'.[19] We could if materialism were true, but if it is not, then the time-slice may just be the commitment of theories designed to make sense of our experiences, but which do not include them. Given that only materialists have ever supposed that physical science is even *meant* to include experience, rather than 'just' all of physical reality (is that not enough?), this latter outcome would not be surprising.

Considerations of simplicity can only count in favour of materialism after it has been established as a viable metaphysical theory, able to include everything that exists. With this established, we would then need some reason to think it is true – and *then* considerations of ontological parsimony might be appealed to. So, Quine has it the wrong way around. Moreover, even if we accept that parsimony is a principle of evidence, it would still only count in materialism's favour against a non-monist metaphysic, such as dualism, which appears to be the only alternative Quine had in mind.[20] Materialism and idealism are equally monistic. So, all Quine has provided is a highly contentious reason for thinking that if materialism and dualism are both viable theories, then materialism is the more likely of these to be true.

Perhaps a justification for Quine's materialism can be found in his requirement that the idealized language be extensional; this is a central theme in Quine, again inherited from the positivists. Since non-extensional language is sometimes used to describe mental states, this explains some of his antipathy to them. Thus, for example, if we take a sentence like, 'John believes Venus is a planet', we cannot automatically read off an ontological commitment to an object called 'Venus' from the truth of the sentence, since whether it is true depends not just on Venus, but also on how Venus is referred to. John may not believe that Hesperus is a planet, since he may not know that Hesperus and Venus are the same. And he may believe that Nibiru is planet, which it is, but a fictional one.

This is exactly what you would expect on the idealist picture, however, since the sentence is not simply referring to Venus, but to Venus in the context of John's thought. The language is as indirect about Venus as our access to it in thought and experience seems to be. This will not suggest an imprecision in language, such that a difference in words is not being correlated with a difference in the world, unless you are already presupposing the materialist view that there are only physical things to correlate words with. If you are not, it will suggest nothing more than that John's thoughts about Venus, Hesperus and Nibiru are all different. And in any case, as Rorty once pointed out, such sentences can always be artificially reconstructed to make them extensional, if so desired.[21]

I have failed to locate the game-changing argument for materialism in Quine, then, one which might have rationally persuaded his profession to turn to materialism. As far as I can see, Quine simply stopped believing that phenomenalism had any epistemological claim on his own systematising project, and so stepped into the scientific description of the world to continue his work from there. Once there, he found that the scientific description does not include experience. But rather than seeing this for what it is, namely an excellent reason to reject materialism, he assumed that physical descriptions *must* include experiences in some way. And the reason that he thought they must is that by this point he was already a materialist.

Smart's 'Sensations and Brain Processes' (1959) was the breakthrough article for materialism, so perhaps that is where the argumentative action occurred. It did, in the sense that Smart understood the contours of the debate, and hence tried to provide a materialist theory of consciousness to show that the materialist's candidate for ultimate reality can explain the idealist's candidate as non-ultimate. But why did he make that choice in the first place? What persuaded him that materialism was worth the effort? He tells us in the second paragraph of the article and then the third begins: 'The above is largely a confession of faith'.[22]

In this crucial paragraph, Smart follows Quine by saying that parsimony is his main motivation. But he is more candid in saying why he thinks materialism is the simplest metaphysic: because it is simpler than Cartesian dualism.[23] The understanding of Cartesian dualism he subsequently displays is exactly that presented in my intuitive case for materialism: the objectified parody of the position, in which we are invited to lose awareness of our own minds while surveying the objective world, and consequently find it unfathomable that the dualist believes in wispy objects floating around in that world. On the basis of this misunderstanding, then, Smart argues that the materialist need only account for the physical world, whereas the dualist has the extra task of accounting for non-physical occupants of the physical world – so the materialist's task is simpler. But at best, to repeat what I said about Quine's original rendition, this could only show that if materialism and dualism can do the same explanatory work, then materialism has the upper hand.

Smart says that, 'There does seem to be, so far as science is concerned, nothing in the world but increasingly complex arrangements of physical constituents. All except for one place: in consciousness.' This is what you would expect on any anti-materialist picture, but Smart does not see this as evidence against his theory, but rather a cause to be suspicious. He does not see cause for philosophical reflection on the descriptive scope of science, but rather a dubious challenge to its explanatory power which needs to be overcome through philosophical analysis. As he goes on to say, 'That everything should be explicable in terms of physics [...] except the occurrence of sensations seems to me to be frankly unbelievable'. In other words, his materialism makes it impossible for him to believe that sensations cannot be explained by physics.

In accordance with his self-confessed faith in science, Smart instinctively looks to the future: 'Certainly we are pretty sure in the future to come across new ultimate laws of a novel type, but I expect them to relate simple constituents: for example, whatever ultimate particles are then in vogue.' Thus, it does not really matter what these 'ultimate particles' are – the metaphysical task of determining the ultimate nature of reality has been reduced to that of urging trust in whatever scientists happen to be saying at the moment.

If we combine Quine and Smart, then, the best case for materialism I am able to detect so far is that *if* you think a materialist account of consciousness is viable, and *if* you are a strong believer in Ockham's Razor, and *if* you think the only viable alternative to materialism is Cartesian dualism, then you have reason to believe that materialism is true. I doubt many contemporary materialists would buy into all three ifs. However, since the viability of a materialist account of consciousness is a prerequisite for the argument to go through, as you would expect, then you have to ask why materialists since Smart have worked on the assumption that a materialist account of consciousness *must* be viable (if this one does not work, move onto the next, etc.). Perhaps because they do not think Cartesian dualism is viable. But that would make the appeal to Ockham's Razor irrelevant. And if the case for materialism boils down to dissatisfaction with Cartesian dualism, there is no case worth mentioning. Many philosophers within Descartes' own lifetime were dissatisfied with his theory. But seeing significant insight within it, his successors went on to develop some of the ideas – this transpired over centuries.

Henry Ford and Rosenberg's view that 'history is bunk' seems to have played a crucial role in the re-emergence of materialism.[24]

We have yet to consider the most famous argument for materialism, however: the causal closure argument. This is the one which would immediately spring to the mind of any contemporary materialist if asked to provide a direct justification for their position. I suspect it is the only one that most philosophers, materialist or not, could even think of. Considerations about how non-physical minds could causally integrate with the physical world were clearly a key driver in Smart's thinking during the key period, even if the notion of a distinctive argument for materialism along these lines had yet to be formulated. It is what Smart has in mind when, in a later paper, he noted that, 'No enzyme can catalyse the production of spook!'[25]

However, the most famous argument for materialism is not an argument for materialism: it is an argument against Cartesian dualism. It says that since the physical world is a causally closed system, if there is anything non-physical in addition to the physical, then it could not interact with the physical world. So, if the argument works, we cannot have non-physical minds which causally interact with our physical bodies, as Descartes thought that we did. If successful, the argument would show that dualists cannot explain the appearance of interaction between our minds and bodies with causation, and so must find another way to explain this appearance. This is the lesson Descartes' immediate successors, such as Malebranche, took from the difficulty of understanding how radically different substances could interact; the specific issue about causal closure is simply an application of this overarching metaphysical concern. Malebranche, like others at the time, appealed to God's omnipotence. When the theological underpinnings of this kind of response became less popular in the nineteenth century, philosophers experimented with epiphenomenalism, according to which the mind is caused by the body, but unable to affect it.

These are some of the lessons which dualists might take from the argument; but only if it works, of course. Causal closure is very much an article of faith, since there are plenty of effects which physical science does not know the causes for; as I write, for example, scientists are trying to discover why the centre of the sun is not its hottest part. The scientific reason for trying to explain these effects is not because the unknown causes are known to be physical rather than non-physical, but rather because to assume they were non-physical would be to give up on science. But if we convert this methodological principle into the metaphysical assumption that the physical world is a causally closed system, then this does indeed create a problem for Descartes' brand of interactionist dualism. It is illuminating to consider why.

As practically any undergraduate student of philosophy can tell you, Descartes' attempt to explain how an immaterial mind could interact with a physical brain was terrible. For he thought the interaction transpired courtesy of the pineal gland, and since the pineal gland is a physical part of the brain, this obviously misses the significance of the problem, which is that of how *anything* physical could interact with something non-physical.

The reason Descartes suggested this, however, is that it made sense within the science of his time – within his own physics, in fact, since he was a practicing scientist as well as a philosopher. Within Cartesian physics, momentum could be conserved even if a body changed direction.[26] So Descartes reasoned that there was scope for an immaterial mind to change the direction of the 'animal spirits', which he thought of as liquids flowing through the nerves, and to thereby affect the body while conserving momentum in the physical system. The law of conservation of energy was developed after his time. Given the scientific knowledge of the brain available to him, then, the pineal gland seemed the most likely candidate for where this interaction takes place. So, Descartes appealed to the science of his time, as materialists recommend that we always should. And because of this, he has been criticised ever since for missing the philosophical point.

If there are materialists in the future who read those of today, they will no doubt find similar embarrassments: philosophical speculation reliant upon what will then be obsolete quantum theory, neuroscience, etc. If there is a general lesson to be learned from the causal closure argument, then, it is to keep philosophical reflection firmly rooted in something more lasting than the best science of the day: our common experience and reason. Future scientists may discover something completely inexplicable that resides exclusively within human brains, and that would not be a good reason for a renaissance of Cartesian dualism. It would make the idea that we have immaterial minds which affect our brains seem more likely, but it would not make the general idea of such an interaction any more philosophically satisfactory.

In my search for a rational justification for the twentieth century renaissance of materialism within the academic discipline, then, I have considered: (1) the most influential materialist philosopher of the era; (2) the most influential materialist article of the era – the one which spawned decades of attempts to produce a materialist account of consciousness; and (3) the best-known 'argument for materialism' of all time. Are there other arguments I have not considered? Maybe there are, but not a well-known one, or one which could plausibly be credited as a decisive factor in establishing the materialist hegemony. Then again, maybe there are not, given that Stoljar, who has written a monograph on contemporary materialism as well as the *Stanford Encyclopedia of Philosophy* entry for 'Physicalism' (these entries are supposed to be state-of-the-art), has only been able to isolate *two* arguments for materialism: an 'impressionist' one, which roughly corresponds to the intuitive case for materialism with which I began this chapter, and the causal closure argument.

If that really is it, then something went very badly wrong when analytic philosophers started taking materialism seriously. There are more arguments for the existence of God – maybe they are useless too, but at least they gave us plenty to think about.

§4. The Let-Down

Perhaps I am asking for too much. I granted from the outset that the main thing a materialist needs to do to establish their position is provide a convincing theory of experience. Perhaps asking for a direct argument is unnecessary. Perhaps materialism simply starts as a hypothesis, an 'article of faith' as Smart put it, and then the theory of experience makes good on that hypothesis. If a theory could be produced which provided genuine insight into how physical activity within certain organic beings amounts to exactly the same thing as a subjective experiential perspective opening up onto the world, then the argument would be as good as over. Strictly speaking this would only show that materialism is viable. But if it started to look as if science really could explain the fact of subjective reality, then map it out with anything like the kind of detail it achieves with objective reality, it would, at that point, be splitting hairs to demand more. The main objection to materialism would have been overcome, and it would seem that science really was capable of describing all of reality without remainder.

So, although there is nothing reasonable about materialists assuming that a materialist account of consciousness *must* be possible, it does seem reasonable to investigate the possibility of such an account. That project has been intensively pursued for over sixty years now. The result has been failure. This is because it has become increasingly clear that materialists must deny that consciousness exists. The so-called 'eliminativist' or 'illusionist' materialists do not like to put it this way, because they want to try to mitigate the patent implausibility, but that is the view, just as the labels suggest it would be.

It is Dennett's view, based on a wealth of evidence from contemporary science – evidence showing that if you look for experience in the objective world, you never find it. It is the view that when we judge that we are having an experience, we are making a false judgement; rather as we might falsely judge that we see a ghost. Experience is a 'user-illusion', as Dennett now puts it.[27] But if the end point of attempting to produce a materialist theory of consciousness is the view that there is no consciousness, then I cannot see how that can be thought to amount to anything other than failure for materialism, but success in the project of trying to find out whether it is viable: we found out that it is not. Imagine your project was to give a scientific explanation of light without mentioning electromagnetic radiation. If your conclusion was that there is no light, only false judgements about light, then that would just be a misleading way of saying that it cannot be done.

The story of this project takes something like the following form. In the beginning was the Place/Smart theory. It held that experiences are physical processes in the brain. Philosophically, at least, this was neuroscientist Francis Crick's 'astonishing hypothesis', as he called it in the title of a book, which he made in the 1990s as part of a successful manifesto for scientists to take consciousness seriously.[28] The philosophical novelty of the Place/Smart theory was its explanation of why experiences do not seem to be brain processes. And the explanation was that our ways of thinking about experiences are distinct from our ways of thinking about brain states, even though we are thinking about the same things. In other words, we have two sets of concepts (mental and physical), normally isolated within our thinking, but which pick out the same things.

But surely if whenever we take ourselves to be thinking about an experience we are really thinking about a brain state, then we cannot be thinking about the brain state correctly; rather as if we seem to be thinking about a book, but are really thinking about a tree, then we are not thinking about the tree correctly. This conclusion seems especially hard to avoid for the materialist, given that their only reason for saying that thinking about experiences is a different way of thinking about brain states, is to explain why it does not seem as if we are thinking about brain states at all. Since this different way does not remotely suggest to us that we are really thinking about brain states, then, it must be the wrong way.

If we draw this conclusion, then we have arrived at eliminativism immediately. Nothing fits our conception of experience, so what we seem to be thinking about does not really exist; rather as if we thought of trees as if they were books, the books we took ourselves to be thinking about would not really exist. Materialists did in fact arrive at this conclusion very quickly indeed – Smart's breakthrough paper came out in 1959, and it only took until 1963 for Feyerabend to publish a paper which drew the eliminativist conclusion.[29] I think it was the right conclusion because materialism is eliminativist by nature. But the Place/Smart theory held out a hope for this conclusion to be avoided, thereby allowing materialists to continue to think, for the time being, that they could have both materialism and experience.

Place and Smart did this by arguing that when we think of a brain state as an experience, we are not in fact thinking about the brain state incorrectly, only differently. This is because we are thinking about it in a stripped-down, basic kind of way. We are thinking of the brain state in accordance with its causal role, or more broadly, its function. So, when we think about a brain state as a feeling of pain, for instance, we are thinking about it as the kind of state we typically occupy when we undergo bodily damage, like a paper-cut. If this is right, then there is no incompatibility between mental and physical concepts and hence the eliminativist conclusion can be avoided.

But even supposing that when thinking about an experience, we do always think about its role, this is evidently not *all* we are thinking about. We are also thinking of the experience as something subjective with experiential character – what it feels like to you, basically. I can think of my period of sleep last night in terms of its function of keeping me healthy, but if I recall any dreams, there will also be subjective, experiential character for me to think about. Thinking of an experience as subjective is rarely an explicit, conscious theme for us, except in philosophy, and the same could be said of thinking of the role it plays in your life. But nevertheless, when you think about a pain you are having, you do not think of it as something transmitted throughout the world – as something objective like a tree, which stands for all to behold. So, it is implicit in the way we think about experiences that they are subjective: there for us alone. Thinking about the experiential character of our own experiences, on the other hand, is always fully explicit. You can hardly think about the pain you are in without thinking about what it feels like. Since brain states are not subjective and do not have experiential character, then, if thinking about brain states as experiences means thinking about them as subjective and experiential, then this must be an erroneous way of thinking about brain states.

This crucial point was not lost on either Place or Smart; although it was soon overlooked in the ensuing debate. For both did explicitly reject subjective and experiential ways of thinking about experiences; the word they used was 'phenomenal' – they rejected 'phenomenal' properties and the 'phenomenal' language used to describe them.[30] What they did not realise, however, was that this means embracing eliminativism. For if we should not be thinking about our brain states as subjective and experiential – if this is an erroneous way of thinking about them, as they agreed it was – then we should not be thinking of them as experiences at all. So, there are no experiences, only brain states. Place and Smart did not see it this way, because they seemed to think it was enough, in order to secure a distinctively mental way of thinking about brain states, to point out that thinking about our brain states only in terms of their causal roles is distinct from thinking of their physical nature as brain states. But that cannot be right, because we can think about the causal roles of both a brain state and an experience, and these will be distinct thoughts, even if the causal roles are the same. To distinguish these thoughts, you must bring back the subjective experiential character.

The flaw to any materialist attempt to account for consciousness was at the surface from the outset. For if an account of subjective experiential character cannot be given in objective terms – one which allows us to understand how thinking of a brain state as an experience could amount to thinking about objective features of that brain state – then these essential features of experiences must be disavowed. And that means experience itself being disavowed – written-off as a non-entity which confused thinking projects upon the world. Given how implausible the eliminativist conclusion is, however, and how damaging it is for materialism, only really radical thinkers like Feyerabend and Rorty were prepared to embrace it in the early days. Since the rest were not, the story moved onto its second stage.

The Place/Smart theory had not explained how subjective experiential character could be possessed by brain states. The hope was that we could abandon the only distinctive conception of experience we have by finding another reason to label brain states as 'experiences'. But it was an unconvincing reason. As a consequence, philosophers continued to rely upon our ordinary conception of experience, realising that it bears practically no relation to our conceptions of physical brain states. This realisation, and the wider situation in which philosophers were becoming increasingly interested in computing and AI, led to functionalism. Since computation is not tied to any specific physical implementation, and since identifying experiences with human brain states seemed to rule out anything non-human having those experiences – dogs feeling pain, for instance – the idea now arose that experiences are not physical, but rather functional states of our brains. This might sound bad for materialism, but not if brain states are performing the functions, rather as the physical innards of a machine perform its functions.

But functions are abstract and experience is concrete. Experiences and physical objects provide our paradigms of concreteness, especially when the two go together. When idealism toyed with eliminativism in the opposite direction, through Berkeley's view that there are no physical objects, only experiences of physical objects, then Dr. Johnson, according to the famous anecdote, kicked a rock to refute it. Rocks are paradigmatically concrete realities. Johnson's confidence in the concrete reality of the rock, however, would have sapped somewhat if he had not experienced his foot rebounding from it, since experiences are paradigmatically concrete realities too. Numbers, on the other hand, are abstract – intellectual abstractions from the world we experience – which is why we do not experience them. So, there was never any prospect of experiences *being* functional; to think otherwise is to make what Ryle called a 'category mistake'.

So, functionalism, construed as a materialist proposal, only makes sense as the idea that when we think about an apparently concrete experience, then we are really thinking about a concrete brain state abstractly; in accordance with the functional role it plays. But then, this is just the Place/Smart theory all over again; it is unsurprising that Smart, quite rightly, failed to see the difference.[31] Moreover, exactly because our conception of an experience is of something concrete, it is clear that if we are indeed thinking about a brain state, as the materialist insists that we are, then we are not *only* thinking about its functional role. And that leads us back to the central flaw with the materialist project once more: if we are thinking about the concrete reality of the brain state as possessed of subjective experiential character – which we must be if we are thinking of it as an experience – then unless objective science can account for this character, we have it all wrong. There are no experiences, only brain states.

The fact that materialism cannot explain why physical states of *any* kind should have subjective, experiential character is the real reason we are ready to suppose that dogs, aliens or robots could have experiences, regardless of how physically dissimilar their internal states might be from our own. The reason is not, as functionalists have supposed, that our conception of experience is somehow entirely abstract, such that we think of it as something which could be realised by a variety of different kinds of physical state. Nevertheless, working on the assumption that materialism must be true, philosophers came to think that experience must be a higher-order physical feature of brains, and hence that a materialist account of experience would have to be 'non-reductive'. Just as liquidity is not a feature of individual molecules, but rather a higher-order property that results when they group together, so experience, it was thought, must be some kind of higher-order property. Since innumerable different arrangements of molecules could give rise to a higher-order property like liquidity, this seemed to explain why experiences could not be tied to any particular physical kinds and thereby reduced to brain states.

So, decades were spent working out how best to formulate a non-reductive version of materialism, using concepts such as 'emergence' or 'supervenience'; it still goes on, with 'grounding' being the current buzz-word. But this was to jump the gun, because unless materialism is a viable theory – which is what the production of a materialist account of experience was supposed to be establishing – then there is no point in polishing the formulation. The move to a non-reductive version of materialism made no progress on this agenda. That the problem had simply been reformulated became clear when philosophers in the 1980s started talking about an 'explanatory gap', an idea formulated explicitly on the assumption that experience is a higher-order physical property.[32] But to close the gap, you still needed to explain experience in physical terms. That is what materialism always needed to do. Reconceiving the task as that of how higher-order properties which are experiential can be realised by lower-order properties which are not, made no headway on this agenda.

The third stage in the story began when some philosophers, who were not overly concerned about whether a physical account of experience should be reductive or non-reductive, returned to the original Place/Smart strategy of trying to explain why it does not seem as if we are thinking about brain states when we think about experiences. So, in the 1990s, these philosophers provided a new account of how we might be able to conceptualize brain states as experiences in a way which neither conflicts with materialism nor reveals the physical nature of the brain states being conceptualized.[33] Of course, this strategy would be unnecessary if materialists could explain why thinking about experiences *does* seem like thinking about brain states, that is, why it seems that way with the benefit of a materialist account of subjective experiential character. But since materialists have never had the remotest clue how to produce such an account, their options are limited.

When the original strategy was returned to this time, however, there was a determination not to renounce subjective experiential character. For by this time, in the 1990s, materialists had long since been bombarded by thought experiments that reminded everybody what the ordinary concept of an experience amounts to. The most famous two, about the experiences of bats, and about stepping into a world of colour after life in a black-and-white room, had already been analysed to death; with materialists never giving an inch, of course.[34] But they clearly had an effect, given that materialists now focused their hopes on the so-called 'Phenomenal Concept Strategy'. 'Phenomenal' concepts, remember, were exactly what Place and Smart had rejected.

In its essentials it was not much different from the Place/Smart original, however. Experiences were still brain states, and the reason thinking about experiences does not seem like thinking about brain states was still that we were thinking about brain states in a stripped-down, basic kind of way – a way which was most definitely *not* incorrect (for to admit that would be to embrace eliminativism). The main difference was that the Phenomenal Concept Strategy insisted that thinking about brain states in this stripped-down way was the same thing as thinking about their subjective experiential character. The idea was that when we experience pain, or a vivid blue, for instance, we are inwardly pointing to certain kinds of neural structure. But since our inwardly pointing concepts are conceptually isolated from our physical ones, and only the physical ones give us any descriptive purchase on the nature of the brain structures being conceptualized, our concepts of experiences conceal their own identity as brain states. Experiential concepts simply provide the ability to distinguish our own personal brain states from each other, within an autonomous, conceptually isolated sphere of thought. At around the same time as this approach became dominant, and by following a similar line of thought, Colin McGinn came to the quite different conclusion that since our concepts of experiences are cognitively isolated from our concepts of brain states, the mind-body problem could never be solved.[35]

So, the new strategy was to leave experiential concepts as they are, but to show that they are not erroneous, as opposed to the Place/Smart strategy of abandoning them in favour of non-erroneous alternatives. But eliminativism was still not avoided. To see this, imagine an experience of a circular patch of blue, against a broadly black, but undulating and complex background; you can probably conjure up one of those for yourself. Now suppose we take the Phenomenal Concept Strategy of saying that I have not misconceived my brain state, but rather conceived it in a manner that has nothing to do with physical or functional concepts and which basically just amounts to mentally pointing at it. Yet here I am, occupying my own subjective viewpoint, and apparently confronted by a circular patch of blue. Even if my concept of this patch does indeed blindly point to a neural structure, that is evidently not *all* it does, since I cannot help thinking of it as something with subjective experiential character, and hence *prima facie* at odds with materialism. If this were not true, then materialists would have had no problem to begin with.

So, my positive conception of the patch must still be renounced, just for a new reason. I do not renounce it because the brain state is not blue, as Place, Smart and the eliminativists do, because on this proposal it actually is – in the metaphysically trivial sense that it is the kind of brain state that we introspectively conceptualize as blue. Rather I renounce my enduring inclination to take my conceptualization as committing me to the existence of something properly described as a subjective patch of blue; I renounce my natural conceptualization of the conceptualization, if you like. I get to keep the subjective patch of blue, but only on the condition that if I meant anything more than 'that' when describing it as a subjective patch of blue, as I obviously did, then I must renounce my description as being applicable to nothing. And that just brings us back to eliminativism. No matter how complex materialists make their theories, they can never escape the essentially eliminativist tenor of materialism itself.

Another popular materialist proposal of this same period was that to have an experience is to represent the world in a certain, objectively specifiable way; this was sometimes combined with the view that experience should not be confined to brain states, but is rather to be found in the subject's wider involvements with the physical world.$_{36}$ But it makes little difference what states, processes or widespread systematic interactions you seek to identify experience with, if the objective description gives you no reason to think that something subjective is thereby being described. Imagine an unconscious AI 'reading' one of these theories. Would it learn why we are different from it? Of course not: the eliminativist position that we are not different would be the conclusion this machine would draw. A human being reading one of these theories, on the other hand, and failing to gain any insight into how objective conditions could possibly account for subjective ones, would be better off drawing the conclusion that we should not try to think about experience objectively – that this is the wrong way to think about experience – and hence that materialism is irreparable.

The fourth stage of the story will also, I suspect, be its last: the homecoming of eliminativism. Dennett has been around for most of the story, and it now looks as if he was right all along, as his mind increasingly turns to where our technological advances are taking us – hopefully not from *Bach back to bacteria*, as he encapsulates this concern in the title of a recent book.[37] For decades, he has tirelessly shown us that if you take an objective, scientific approach to experience, then you discover more and more reasons to think that experience, as we ordinarily think of it, cannot really exist. Given that materialism tells us to defer to science, and has always looked like an essentially eliminativist position, I think he was right all along. Of course, the moderates who want to be materialists without denying the existence of experience may still come up with something new. It is not hard to see why moderate materialists want this: denying everybody else's experience – solipsism – is paradigmatically crazy, so why would denying your own into the bargain be thought to improve the situation? But if materialism was only ever a hypothesis in the first place, and we have discovered that its natural direction of travel is eliminativist, I see no good reason to continue trying to resist that direction of travel, rather than simply learning from it.

To recap, then, materialism needed to provide an account of consciousness to show that it is a viable metaphysic. It needed to demonstrate its viability against the basic objection that it accounts for the physical world we experience, but not experience itself. Given that materialism's main focus has been consciousness ever since its renaissance, when philosophy of mind suddenly became central to metaphysics, this was clearly not missed – there is little more disingenuous than a materialist trying to play down the significance of consciousness, as some try to these days. But if materialism must deny the existence of consciousness, this simply confirms the original suspicion that it is not viable.

Moderate materialists could, at this point, retreat to mere faith that science will one day explain consciousness. But they cannot pretend there is anything remotely reasonable about this, given the lack of direct arguments for materialism. For if materialists cannot even imagine the vague contours such a solution might possess, as they evidently cannot, they have no reason to think science ever will provide this materialism-validating explanation. The long slide of materialist metaphysics into eliminativism, on the other hand, gives us good reason to think it will never happen. To think this is not to underestimate science; it is not to suggest shortcomings in the scientific method; it is not to despair of our cognitive abilities; it is not to prophesize a coming intellectual revolution. It is so much simpler than all that. It is simply to believe, on the basis of philosophical reflection on the recent history of materialism – as well as broader historical reflections – that the very idea of an objective account of subjective experience is a false explanatory task which has been imposed on science by the false metaphysic of materialism.

§5. The Loop

In order to give materialism one last shot, perhaps we could try to see its recent history as rational progress towards the realisation that experience does not exist. This is somewhat perverse, since materialism's ability to explain experience was supposed to have been its saving grace, and for there to have been rational progress towards the view that experience does not exist we would have needed a rational basis for believing that materialism is true from the outset. But maybe the materialists guessed right and discovered something important as a consequence. So, let us consider eliminativism on its own merits. I shall follow the Dennett-style 'false judgement' account, since it is the only one I really understand.[38]

Now, as we normally think about it, we know about physical objects, like trees, because we experience them. We know what experiences we are having and this allows us to recognise things like trees. If eliminitivism is true, however, then the situation must rather be that when we *think* we recognise an experience (you cannot recognise something which is not there), then we are making false judgements about the existence and nature of something which is not really there at all.

But eliminitivism is a recent, controversial view, which has obviously not affected the development of our sciences. Observation is the basis of science. So, when scientists were drawing up the theories of the world we now rely upon in so many ways, they must have thought they were experiencing certain observable patterns in nature; patterns they could theorise about. So, for example, they found that whenever they burned sodium, they experienced an orange glow. Their experiences of the orange glows were false judgements, the eliminativist supposes. But our understanding of what the orange glow really is, the actual physical process transpiring in the flame of the Bunsen burner, is an understanding of the kind of thing that causes our experiences of orange glows. So, if experiences do not exist, it seems to follow that our understanding of sodium, and everything else in the physical world, is an understanding of the causes of various non-entities.

Suppose you watch a CGI fantasy film about a world of wizards and dragons. If that was your only evidence about the nature of reality, then your account of reality would, of course, include wizards and dragons. Would it make sense for this account to lead you to mistrust your initial evidence? Not in a completely wholesale fashion. For to come to believe that the whole film had simply been inducing in you false judgements about non-entities, would be to allow your views about the world of wizards and dragons to persuade you that you never had any reason to believe in that world in the first place. But, of course, if you never had any reason to believe in it in the first place, then it cannot have provided you with good reason to believe that you never had any reason to believe in it in the first place. So it is simply incoherent to suppose that from completely illusory evidence, you could draw up a picture of reality which reveals that evidence to be completely illusory. If the evidence is completely illusory, you could never know this, so you are rationally compelled to regard the evidence as at least largely trustworthy.

So, if there are no experiences, only false judgements about non-entities, then experience is not the evidential basis of our understanding of the world; our false judgements about non-entities provide this basis. We drew up our picture of the world by working out the kind of things that could cause these non-entities; non-entities which we thought of as experiences of trees, of sodium burning, of readings on a Geiger counter, etc. In that case, our physical understanding of the world is a reflection of the non-entities which provided its evidential basis. But then it would be incoherent to rely upon this physical understanding to cast wholesale doubt upon its own evidential basis.

Limited distrust is coherent, since it evidences general trust. For example, we have learnt that we can experience physically different kinds of surface as equally red; that physically quite different things can cause us to have the same kind of red experience. So trust in the ability of our experiences to immediately reveal to us differences between surfaces is limited against a backdrop of general trust that we do indeed experience the shared redness of the two different surfaces, if not the features that make them different. This general trust can be relied upon when seeking the more rarefied, equipment-mediated experiences required to determine what the differences actually are. If there are no experiences at all, however – if everything we have ever been inclined to say about them, even down to our most basic commitment of all, namely that they exist, is to be disavowed – then they can provide no basis for a trustworthy understanding of the world. So, the understanding of the world we have can provide no basis for eliminativism. If eliminativism were true, we could never rationally assert this truth. This is 'the loop'.

As I said earlier, Democritus knew about it. His materialism led him to believe that, 'In reality we know nothing – for truth is in the depths'.[39] So perhaps it was unfair when an ancient commentator wrote that:

> Democritus went wrong in a manner unworthy of himself when he said that in truth only the atoms are existent, all the rest being by custom. For according to your theory, Democritus, not only shall we not be able to discover the truth, we shall not even be able to live, taking no precaution against fire or death . . .[40]

Perhaps that was unfair, because Democritus was seeking peace of mind in a kind of scepticism, rather as Epicurus was later to look to materialism for peace of mind in the face of religious anxieties, inspiring his followers to 'put their faith in a stern limitation of human ambition, a concentration on what little is possible to us here and now', as Midgley puts it.[41] Contemporary materialists, by stark contrast, think science can uncover Democritus' truth 'in the depths' and hence that we can, in principle, know everything. Perhaps ancient materialism was more philosophically sophisticated, then. Ancient materialism was argued for, even though nobody today would want to resurrect those arguments, based, as they were, on *a priori* reasoning about imperceptible matter. The science it inspired has progressed beyond all recognition when it comes to reasoning about imperceptible matter. Materialist philosophy, on the other hand, seems to have regressed. The combination of science moving forwards and philosophy moving backwards is not good for our world.

4

A New Idealism

§1. Seeing the World in a New Light

Materialism should no longer be given the benefit of the doubt, because there is no reason to think science will vindicate materialist philosophy with an objective account of experience, and no reason to think materialist philosophy will make progress in dismantling the concept of experience. There is good reason to think that neither of these things will happen because the expectations are produced by a faulty metaphysic. The realistic scenario is that science will steadily improve its description of the objective correlates of experience, until the point at which technologists can produce AIs that look and act as if they are conscious. This would not settle the debate. It would simply bring it out into the open, where, if we stay on our current trajectory, the discussion would be less rational and more emotionally charged.

The advent of these machines, should it transpire, will persuade many to think of themselves as objective machines, just as materialism recommends. For we will form emotional bonds with them, especially when they start to 'care' for our children and elderly; children (and adults) form strong emotional attachments to teddy bears, so they will obviously form much stronger attachments to things which look and act like real, caring people. However these emotional bonds will present an obstacle to the economic forces which brought them into our world in the first place: an obstacle to our treating them as slaves. Our new gladiatorial games would seem cruel. So, we will want to think they are fundamentally different from us, in order that we can use and abuse them. Just as the production of the machines will generate bad, emotive reasons to think they are like us, it will generate equally bad, emotive reasons to think they are not. If greater philosophical awareness could arise in this world, to which I think materialism is the main philosophical obstacle, then we might decide not to get ourselves into this pickle in the first place; but if we do, we will handle it more rationally.

What we need is a metaphysical Gestalt shift. Otherwise, as Maurice Merleau-Ponty once put it, 'the world will close in over itself, and, except for what within us thinks and builds science, that impartial spectator that inhabits us, we will have become parts or moments of the Great Object'.[1] To avoid this incoherence – and the erratic behaviour which incoherent thinking can prompt – we need to see that philosophical debates matter, because they can inspire technological developments that change our lives and concern the attitudes we should take to these developments. We need to see that materialism is not an extension of ordinary objective thinking about the world, but rather the most significant philosophical challenge it faces in our day. We need to see that mathematical physics does not describe the world we see and touch, but is rather a means of predicting and controlling our experiences of that world. We need to see that we have no reason to believe that a reality which exactly and exhaustively reflected our physical theories would include experience, and hence no reason to think it would be *our* reality. We need metaphysical beliefs rooted in the concrete reality of experience, rather than the abstract projections of mathematics. In short, we need idealism.

The history of philosophy has produced many varieties of idealism. The version I shall argue for is the one I called 'The Transcendent Hypothesis' in my previous book, *Philosophy in a Meaningless Life*. In that book, I tried to develop some of the details of the position over the course of three chapters.[2] My aim here is different. It is to describe its essential core and indicate its main social significance, before proceeding to argue for it. The five arguments I will present are direct arguments, the kind which were never supplied for materialism. One of the main lessons I take from the recent history of materialism, as discussed in the previous chapter, is that having good reasons to take a position seriously in the first place is a task which must take priority over working out the details. Never start with a lacuna.

Now in the previous book, I said that the Transcendent Hypothesis was not idealism, and I also denied it was realism, preferring to portray it as somewhere in-between the two.[3] I gave two reasons for declining the 'idealism' label: the first was that the theory embodies a certain kind of scepticism about the concept of experience, and the second that it does not rule out the possibility of there being non-experiential elements to reality which we do not know. But although these are not stereotypically idealist ideas, neither are they definitively non-idealist. What does now strike me as definitive is that the theory makes two distinctive claims which originated in the historical traditions of metaphysical idealism, both ancient and modern, and which remain indelibly linked to them. These are the claims that ultimate reality is transcendent and that it is experiential. So, I changed my mind about the label.

Idealism is not the only alternative to materialism, of course. Currently more popular options are dualism and panpsychism. But just as I will not attempt to justify my favoured form of idealism over others, neither will I make any concerted effort to undermine non-idealist alternatives to materialism. I think a certain solidarity is currently called for among anti-materialists, since we are all trying to place the world in a new and more philosophical light. Nevertheless, I will at least indicate why I favour idealism over dualism and panpsychism, since the reason will be very significant when I go on to outline the position.

Dualism holds that ultimate reality is divided between material reality and the immaterial reality of mind – that it is a combination of the two. Panpsychism holds that ultimate reality is experiential, with experience to be found throughout the physical world; in electrons, for instance. What is wrong with both, I think, is that they leave the materialist's metaphysical conception of the physical world in place. And the reason they do this is that neither gets fully to grips with what it means to think subjectively about experience.

The dualist need not conceive the material world any differently from the materialist, so long as they insist that it does not exhaust the ultimate reality, which, on their view, also incorporates the mental. And the panpsychist also thinks the materialist's conception of the material world is not so much wrong as incomplete, albeit in a different way. According to the leading contemporary version of this view, Philip Goff's 'Russellian Monism', the idea is that the abstract conception of the physical world provided by 'the mathematico-nomic vocabulary of physics', as Goff puts it, tells us about the structure of the world, but not what fills out that structure.[4] It provides us with an extrinsic characterization of the physical world, but leaves us in the dark as to its intrinsic nature, which is experiential. So just as with dualism, but in a different way, materialism is not so much wrong for what it says, as for what it does not.

The reason dualism and panpsychism concede this much to materialism, I think, is that they fail to take on board the distinction between two importantly different conceptions of experience we have. This distinction is made most clearly by J.J. Valberg, although its presence is discernible in many important philosophers, such as Kant, Wittgenstein and Sartre. I shall follow Valberg's terminology by distinguishing between the 'phenomenal' and 'horizontal' conceptions of experience.[5]

When we think about experience phenomenally, we think of it as something that appears to us – a 'phenomenon' is an appearance. So, if you feel pain and think about *that* feeling, then you are thinking phenomenally about your pain. To think about your experience horizontally, on the other hand, is to think about the whole thing – but not as one big phenomenon, nor lots of little phenomena stuck together. For it is not to think of experience phenomenally at all. You no longer think of experience as something you might focus your attention upon, but as, at a minimum, the boundaries of the experiential field itself, the horizon of experience. The horizon is not a something to experientially focus your attention on, but a nothing within which all such acts of focusing become possible. It is an experiential perspective on reality which you recognise not by experiencing it, which is impossible, but by recognising that you experience other things. Objects like trees and rocks do not only exist, they also appear, and that is because people experience them. People experience them within experiential horizons which they cannot experience. I shall have more to say about this when I outline idealism.

Now although dualism and panpsychism are serious about subjective thinking, by operating exclusively with the phenomenal conception they are forced to conceive subjectivity as a special kind of property. Experiences, whether conceived as events, objects, or properties of substances, must either have this property or be such a property. Having or being this kind of property confers experiential status, as opposed to the horizonal account, according to which existence within a horizon does. Existence within a horizon can be conceived as a property too, of course, but when explicitly dealing with subjective thought, this encourages us to forget that it not a property you experience, but rather one you reason to on the basis of what you can experience, so extra care must be taken when using the language of 'properties' in this context.

In order to make sense of subjective properties, a conception of the reality they belong to is needed. So, we could say that subjective properties characterize a subjective reality (the immaterial reality of mind) or an objective reality (that some or all objective things have subjective properties). Different types of dualism and panpsychism will go in different directions at this juncture. But whichever way they go, they will have still inserted subjectivity into the world we each find within a horizon of conscious experience, and so will have inserted it alongside the objective world. Neither position is motivated by the thought that subjective experience and the objective world are *both* found within conscious horizons, and that this is a reason to question their ultimate reality – so neither finds the materialist's conception of the objective world problematic in and of itself.

The result is that dualism and panpsychism raise the traditional mind-body perplexities associated with trying to explain how our reality could be one in which subjective and objective properties co-exist. The problem the dualist faces if they try to explain this causally is notorious: causal closure. Notoriety does not make it a good argument, but it is indicative of the kind of problems that can arise when trying to work with the materialist's conception of reality, designed, as it was, to be all-encompassing. And the situation is not obviously any better for the panpsychist. On this view, an experience of a tree, for instance, obviously cannot have the extrinsic, objective properties of a tree, since the experience belongs to the person having it, not the tree. So, like a materialist, the panpsychist will naturally think the experience has the objective properties of a physical state of the perceiver, such as a brain state, and then explain that state's experiential status through its also possessing an intrinsic, subjective nature. But then if the tree has experiential properties as well as the brain state – in accordance with the premise that everything has an intrinsic nature –

then you wonder what these might be, or how anyone (apart from the tree) could possibly know about them, or be confident in reasoning to their existence. If they are not the same experiential properties possessed by our brain states when we experience the tree – the greenness of the leaves, the roughness of the bark – then it is hard to see why we should believe in them, and yet if they are the same, the motivation for doubling-up in this manner seems dubious. Dualists and panpsychists have sophisticated responses to these kinds of concern, of course, but I think the need to provide them is generated by overlooking an essential element to subjective thought.[6]

§2. An Outline of Idealism, and of its Consequences

The key insight provided by thinking of experience as a horizon, according to idealism, is that consciousness enacts a distinction between different levels of reality. This is not in the sense of a difference between microscopic and macroscopic levels – with the former grounding the latter, or the latter supervening on the former, as is extensively discussed in materialist philosophy. Rather, it is the difference between the phenomenal reality to be found within a horizon of consciousness, and the ultimate reality that transcends that horizon. Since we are trying to develop subjective thought beyond common sense, rather as science develops objective thought beyond common sense, this distinction must be understood phenomenologically, that is, in terms of how things seem to us. The aim is to find a metaphysical understanding which leaves ordinary subjective and objective thought untouched, but which allows us to go beyond them to satisfy our philosophical interests. For unless everyday subjective and objective thought are respected, the understanding will be hard to believe and is liable to be compartmentalized away from life as a separate thing to think about; as only *supposed* to be about the life you know. That is where materialism goes wrong: by trying to negate subjective thought, and by rendering everyday objective thought dubious by interpreting science's extension of it metaphysically.

So, the key distinction is between the phenomenal reality found within a horizon of consciousness, and the ultimate reality that transcends that horizon, and this is to be understood in terms of how things seem to us. As such, let us begin with an example in which how things can seem to us brings the existence of a horizon clearly within our purview. The example concerns dreaming, the same phenomenon Descartes used to bring philosophical reflection into the modern era.

The process of dreaming is one in which you have experiences which, rather than opening you up to the objective world, cut you off from it. For when you are dreaming, the objective reality in which your eyes are closed and your head is on a pillow does not appear within the horizon of your experience. You can wander far and wide in that dream, hop on an imaginary spaceship, perhaps, but you will never find that pillow. If, in the course of your travels, you found yourself gazing down on your sleeping body, in a room that looked exactly like your bedroom, the pillow would still not appear within the horizon of your experience, only the dream's imitation of it would. The objective world where the pillow exists transcends your experience while you are dreaming. When you wake up, however, what was previously the transcendent reality of the objective world will now appear within the horizon of your consciousness. You will now be able to find your pillow experientially appearing to you within that horizon. The leading idea of idealism is that this simple observation reveals to us the general structure of conscious experience: consciousness encloses us within a phenomenal world, from the perspective of which there is always a transcendent reality. From the perspective of waking life, in which both the objective world and subjective experiences appear within our horizons, transcendent reality is *the* transcendent reality. It is the ultimate one to which we can never wake up because of how conscious experience works, its essential split-level structure. A higher awakening from the horizon of ordinary life – after death, perhaps – could only be to a new phenomenal reality, not the ultimate one.[7]

As I stressed in Chapter 2, the ultimate reality, for the idealist, remains within our experience. So, the example of dreaming has the potential to mislead, because the objective world transcendent to the dream cannot appear within the horizon of that dream. The reason for this disanalogy is that a dream is a construction from elements gathered together in waking life, a life in which we understand our experiences with a combination of objective and subjective thought. Our understanding of the pillow is not of something only one person experiences, but of something others can too, because it is a physical object with its own independent nature. As such, the pillow cannot appear within the horizon of the dream, because we cannot understand a dreaming experience to be of that particular pillow and must instead understand it entirely subjectively: as the illusory experience of a sleeper. But that does not mean the dreamer is experientially cut off from ultimate reality, as they are from the objective world. The experience is real, but its ultimate reality transcends not only our understanding of it within the horizon of the dream, but within the horizon of waking life too. As such, the ultimate reality always remains within our experience.

Waking life transcends the horizon of a dream because we cannot understand what appears to us in a dream with the combination of objective and subjective thought that we use to make sense of waking life. Ultimate reality transcends the horizon of waking life, on the other hand, because we cannot understand what appears to us in waking life *without* that combination, except for metaphysically. There is no higher understanding to parallel the higher understanding that transcends a dream; one which trumps the understanding of the dreamer to relegate the dream to mere subjective appearance. The combination of objective and subjective thought is as high as positive understanding gets for humans, which is very high indeed if we leave subjective thought in the background to focus on extending objective thought: science has taken it to dizzying heights and there may be no ceiling. But we cannot transcend the combination, which our way of life and ordinary grasp of reality requires, except through the metaphysical insight that we would need to do so to grasp the ultimate nature of reality. This insight does not trump ordinary understanding when it ascribes to it non-ultimate status, because the new philosophical context it sets up is one in which we can say nothing positive, only reflect on the unaffected context below. In some form, it may be the insight which gave us the notion of metaphysics, and hence ultimate reality, in the first place.

The insight is just that although perfectly suited to our way of life, our combination of objective and subjective thought cannot be brought into alignment to form a unified conception of reality. Billions of human beings each have their own horizon of experience, and from within those horizons, their bodies appear to them in a particularly intimate way which amounts to ownership, along with all their thoughts that range wide over possibility and actuality, feelings that refer back to those bodies and make them of the highest concern to their owners, and perceptions of the same physical world that everyone else can perceive. There are also non-human animals perceiving the physical world from within horizons of experience so incomprehensible to us that some have even doubted whether they exist. Somehow that cacophony of experience must be united within a unitary reality, along with the physical reality they refer to, which we already understand as unitary. If that unified, all-encompassing conception could be provided, we might fully comprehend the ultimate nature of reality. But it cannot be provided, because our understanding is bound to the phenomenal reality that appears to us within our horizons of experience. Our understanding is bound by the structure of consciousness itself, the fact that it encloses us in a phenomenal reality, from the perspective of which ultimate reality is always transcendent. Our positive, non-metaphysical understanding is tied to the phenomenal world it allows us to make ordinary, scientific and otherwise theoretical sense of. The ultimate reality appears to us alright – we call this 'having experiences' – but not in its full splendour, only in a human way.

Ultimate reality transcends experience in the sense that we can only understand it as experiences appearing within horizons and thereby informing us about the physical world, with this understanding being enough for us to recognise its limitations, and hence that it does not allow us to fully comprehend ultimate reality. Ultimate reality cannot appear to us in any other way, a way which might allow us to form a more comprehensive conception, in just the same way that an actual physical object cannot appear within a transcended dream. Idealism does not turn waking life into a dream, as might be thought on the grounds that it portrays waking life as a transcended reality. Rather, it offers higher understanding akin to that acquired when waking from a dream. It allows us to see that our phenomenal reality is transcended, and there is no waking from it. It thereby makes everything bigger, not smaller.

Now subjective and objective thinking are very different. In the case of an objective phenomenon, like an appearance of a rock, objective thought can provide us with a very detailed picture of what rocks themselves must be like to give rise to phenomena of that kind. This is the most detailed and explanatorily powerful mode of thinking at our disposal. But we cannot make detailed sense of experiences. This is because everyday subjective thinking takes place through the analogous employment of objective thought, upon which subjective thought is parasitic. We may lean on everyday objective thought to say that a pain is sharp or throbbing – like a knife or an oscillating wave – or we may use science to go further, by saying it is associated with activity in a certain region of the brain. But we will never achieve the descriptive detail, and hence explanatory power, that any kind of objective thought can provide, for the simple reason that subjective thought takes phenomena *as* phenomena – and thereby points outside itself to a transcendent reality. Objective thought presupposes then forgets phenomena to focus on its account of the conditions in which they arise, but subjective thought resides with them. That is why experience can seem ineffable.

Imagine looking at a field on a bright, sunny day. You have a good visual perspective on it, allowing you to take in its whole massive surface, unimaginably detailed with hundreds of thousands of blades of grass glinting in the sunshine. Objective thought could provide a detailed account of what you are seeing. You might imagine this information being conveyed by a computerized matrix projected over the field, dividing it into tiny regions, and allowing individual plants to be highlighted with clickable links that reveal their angles, measurements and chemical composition. The matrix will descend over the objective phenomenon, but the information will only tell us about the objects which explain it. That is how objective thought makes sense of phenomena, namely by forgetting that they are phenomena.

If you switch to subjective thought, however, you will now think about what it is like for you to see the field. And if you try to describe the difference, you will find yourself tongue-tied. For all you can add to objective thought's story is that this is the experience of somebody with certain objective features viewing that field; somebody with sensory capacities responsive to certain of its features but not others. Even if you try to describe the experience more directly than that, you will still find yourself relying on objective thought, since your experience will now seem to have a certain shape (a top and bottom, for instance), despite the fact that metaphysical reflection reveals that this cannot be right: experiences are not additional occupants of the space of the physical world.

And yet despite the overwhelming dependency of subjective upon objective thought, switching to subjective thought changes how things seem. It will now seem that reality far outstrips any description we can give of it and that no information could possibly convey what you are experiencing as you gaze at that field. The idea that any words could convey that particular, concrete reality to another person who is not actually aware of it is suddenly liable to seem a very odd notion. '"Convey" in what sense?' you might well wonder.[8] The reason is not that experiences are ineffable phenomena, as philosophers have so often supposed, because to describe the phenomena as part of their phenomenal reality, rather than looking beyond them, we must describe the objective world they reveal. For this reason, we can provide just as much detail; much more is needed, in fact, because now we must add the objective details about the perceiver. The real reason for the ineffability intuition is that subjective thinking draws our attention to the fact of appearance within a horizon; an appearance which reality transcends. That we are now explicitly thinking of an experience, rather than the field it reveals to us, draws our attention to the inability of our regular combination of subjective and objective thought to encapsulate the ultimate nature of this reality, which is transcendent. We have realised that our phenomenal reality points beyond itself. After that, subjective thought leaves us cold. But a nod in the right direction is all metaphysics needs.

We become indirectly aware of horizons through the fact that each person's experiences and thoughts are unified into one coherent whole referred back to an active bodily subject: the one seeking new experiences who is the one thinking the thoughts. We also become indirectly aware of horizons through our awareness that they opened from experiential nothingness and will one day close. While they are open, the physical world exists within them, in the sense that it is qualified as phenomena: as appearances of physical objects, at the level of ordinary objective and subjective thought, and, simultaneously, as appearances of ultimate reality, at the level of metaphysics. When they close, the physical world is no longer qualified as appearance, not for the person who died. It no longer exists within that erstwhile horizon, it just exists.

This plain existence is a commitment of both ordinary objective and subjective thought, which are entwined. Our understanding of a physical object is of something that exists whether experienced or not; we cannot understand our experience any other way at this level. Idealism preserves this understanding when extending it to the metaphysical level, because now the experience of the physical object is understood to have always been an appearance of ultimate reality. So, our understanding of the physical object as something with its own independent existence is now seen to have been an understanding of the ultimate reality, which, by definition, has its own independent existence. Our sense of reality has not been challenged, only developed along metaphysical lines. It need not be developed in the materialist, dualist and panpsychist manner of taking it as a statement of metaphysical fact, such that independent physical existence, in part or in whole, is the ultimate nature of reality. If we do that, then either our ordinary sense of the reality of experience will not be respected, as in the materialist vision, or else we will replicate the tensions between ordinary objective and subjective thought at the metaphysical level, as happens with dualism and panpsychism.

Horizons are a commitment of subjective thought: of how things seem to each individual. Experience seems to inform us of a physical world and seems to be unified, and our ordinary understanding is that both these things are true. But since we experience the physical world but cannot experience horizons, recognition of the latter constitutes a theoretical extension of subjective thought, rather as science extends objective thought to take in things we cannot experience. My horizon explains the fact that I experience a world and the unity of my experience around my own point of view on that world; and the same is true for everyone. But although the horizon was theoretically posited, its requirement by subjective thinking itself means it must exist. It is not a phenomenal something binding experience together, although this togetherness is experienced. Nor is it something that is revealed by experience within the same context of existence, as physical objects are. Rather, it is that part of ultimate reality which each of us can individually infer from the fact that we experience a world. And what reflection on our own horizons allows us to see is not only that reality can transcend experience, which objective thought already supposes within a unitary context of existence. It allows us to see that reality can transcend thought too, except as regards our most basic understanding of all, namely that of concrete existence. A horizon limits both experience and thought, thereby setting up a context of existence in which all our understanding is moulded to fit what we can experience, and leaving only the undeniable fact of existence when we try to go beyond that context. We see this in dreams, where existence is the only understanding that survives when we abandon the understanding of the dream to seek the transcendent reality it must possess, and we see it in waking life too.

Idealism only questions ordinary understanding in one respect: whether it can provide a complete account of ultimate reality. Since idealism maintains that it cannot, it extends objective and subjective thought to a new level of understanding while leaving the original exactly as it was before. Materialism also maintains that it cannot, but then impacts so heavily on the original that its moderate advocates are left struggling to find anything to salvage. For idealism, however, if the ultimate nature of reality is transcendent, no new evidence could possibly arise to conflict with the understanding from which metaphysical reflection began – in contrast with the materialist alternative, in which the new evidence continually arising from science is interpreted as conflicting with ordinary understanding. Ordinary understanding has simply been elevated by idealism. And this is its great beauty, namely that it cannot interfere with our ordinary understanding.[9]

The supplementary level of understanding is crucial, however, because it provides a philosophical context from which we can see that one of our most spectacular achievements, namely our ability to extend objective thought with science, will never lead us to the ultimate nature of reality. This has two major consequences. The first is that it removes any possibility of a challenge to science arising from subjective thought, such as through the idea that the existence of consciousness shows that our science cannot be right, which has appealed to so many philosophers and still does. This is socially trivial in our currently unphilosophical world, but it may not always be.

The second, and currently far more important consequence, is that it negates the materialist dogma that seems to have been internalized by so many scientists, according to which science will reveal the ultimate truth about reality; that it will allow us to 'know the mind of God', as Stephen Hawking once put it.[10] Within the new, philosophical perspective opened to us by idealism, according to which the ultimate reality is inaccessible except in its human appearance, we can no longer think it justified for science to discover anything it can. Ultimate reality is not unknowable – it may be endlessly knowable – but it can never be completely known because it is transcendent. Our knowledge of its complete independent nature will always end at the fact that we must believe it has one because we know it exists.

As such, the materialist notion of a disinterested scientific search for as much truth as possible can be seen to be lacking any metaphysical point; so we must look for others. Satisfying a widespread curiosity has a point, but not the curiosity of the individual specialist, team, or scientific community, not except for themselves. It attains a wider point only if it feeds into the promethean project of improving the human lot, whether by design or accident. But in some areas of science, an inadvertent consequence can permanently change things for everyone, in a manner we cannot act or argue against, and thereby seek to rectify once it has already happened – so it is risky. Only a good, publicly debated reason could appropriately justify that kind of risk. *Interested* inquiry, motivated by widespread and public reflection, would be the highest mark of honour for scientific inquiry within an idealist context. Interested inquiry already has this status is in some areas of medical research, for instance, but without widespread public reflection, only approval; this kind of attitude, but combined with reflection, needs to spread to other areas that could foreseeably affect everyone, forevermore. We could learn to think this way without a metaphysical context, of course, but have not done so yet – and the ultimate nature of reality, which is of natural philosophical interest, seems a strange thing to ignore. We have learned to ignore it only because it came to seem of only specialized scientific or religious interest.

To move forward to idealism is not to think the gods have won, because the analogy to the Titanomachy was always a false one. Plato could not have realised, since the distinction between philosophy and science would only emerge long after his death. But we can now see that gods and titans never fight, because the gods occupy a higher context. It is from that higher context that they inspire us to think about how the titans can best help us, and about how we can show them appropriate respect. Materialism and titans have seemed the same, but they never were, and although our failure to notice the difference once provided an impetus to seeking the help of the titans, this conflation has now become unhelpful. The new gods of reason are not the same gods who tortured Prometheus, simply for trying to improve our lot. We can only imagine worshipping gods like that out of fear. But we need no longer fear them, because the new gods and titans taught us how not to believe; or else how to rethink ancient human conceptions of old gods. The titan's contribution was to the technology that removed so many of our fears, and the science which explained others away. Together, the new gods and titans can stop at those fears it is reasonable to address, so that we remain human.

§3. Five Arguments for Idealism

(1) The Argument from Cosmology

Like the traditional Cosmological Argument for the existence of God, this argument focuses on the question of why there is something/everything/anything. In the traditional argument, the existence of God is said to provide the best answer, and this is seen as proof that God exists. The Argument from Cosmology works in a similar manner, but concludes that the best answer is not God, but rather that reality is transcendent. This constitutes an argument for idealism because in order to have a rational opinion on the nature of ultimate reality, the question of why there is something rather than nothing must be addressed. So, if idealism provides a better answer than any other metaphysic, with the others not really providing an answer at all, then we have good reason to believe that idealism is true.

The reason this question must be addressed is that to believe something is the ultimate reality is to believe that no other reality is responsible for its existence; for if it were, it would be a dependent existence and hence not the ultimate one. So since taking up a stance on the nature of ultimate reality requires at least some commitment on this question, it cannot be rational to make a commitment without considering the question. For example, somebody who thinks the physical universe is the ultimate reality cannot think it was created by, or is dependent upon, something else. But if they have not even considered the question of why the physical universe exists – not even to discount it – then it cannot be reasonable to think this.

Now the question of why there is something rather than nothing is a very natural one. We ask it as children, and, if we are lucky, those around us will not be discouraging. But as soon as we try to answer it becomes immediately obvious that we must go beyond objective thought. This is what baffles us, as we find ourselves drawn into philosophy by our natural propensity to reflect on the reality we belong to. The objective thought we usually acquiesce in can no longer satisfy our curiosity, because if we appeal to an objective condition to explain why there is a reality, the objective condition will simply be another part of the reality we want explained.

To appeal to an endless series of objective conditions is just to endlessly defer the question, for every time one objective condition is explained with a prior one, the one before that needs to be explained. To say there has always been a reality is just to say there has always inexplicably been a reality, not to explain anything. No reflective person can find that satisfying. And neither can we sensibly appeal to absolute nothingness as the origin of objective conditions. The Big Bang took place, there is no doubt about that, but it could never settle the metaphysical question. For to state that an original event transpired for no reason is to state that something inexplicable occurred, not to give an explanation. If we come to realise this, and yet still persist with the question, we may find ourselves turning to subjective thought. But this is of no use either if restricted to the idea of phenomenal realities. For construed in this manner, an eternal immaterial mind or a first experience arising from nothing would fail to answer the question for parallel reasons.

At this point, we may be tempted by one of two popular options. The first is that the existence of God answers the question, and the second is that it is simply a brute fact, incapable of further explanation, that reality exists. These were, respectively, the positions taken by Father Copleston and Bertrand Russell in their famous 1948 debate over the existence of God. But both are unsatisfactory and for essentially the same reason.

The first idea is that God is a necessary being, and hence exists according to his own nature. So, the reason God exists is that his nature makes it so, and the reason everything else exists is that God made it so. The problem here, however, is that we have not been told what it is about the nature of God which makes him exist, and as such, the question has not been answered: our attention has simply been redirected from our inability to explain the existence of reality, to our inability to explain the existence of a special part of it. We have no idea what it would mean to have a necessary, self-generating existence. Theologians will say that the nature of God is a mystery, in this respect and others. But that is just to say that the existence of a necessary being *would* explain why there is a reality, *if* we could grasp its nature. Since we cannot, the explanation is never forthcoming, so the philosophy has simply been halted with faith.

The second response, namely that the existence of reality is a brute, inexplicable fact, simply closes off the question at an earlier stage. Thus, rather than saying that the inexplicability of God explains why we should not expect an answer, we simply assert that reality itself is such that we should not expect an answer. This is a weaker response, because at least the nature of God is supposed to be a mystery. But the upshot is the same, namely that we should stop asking. Without an account of what it is about reality which prevents us from answering, however, this is not good enough. To simply assert that either reality or God possesses a nature which prevents us from answering is not to give an account.

So, the task then becomes to explain why not. And now two new options present themselves. The first is to argue that the question is nonsensical; akin to 'what is north of the North Pole?', for instance. The second is an idealist explanation of what it is about reality which precludes its existence being explained. The first closes the door to philosophy but the second opens it – by taking this explanatory shortfall to be highly significant. And this, of course, is exactly how the question seems to us: it seems highly significant, until scientism, religion, or simple impatience shuts down your mind. Idealism provides an appropriate answer. It tells us what it is about reality which precludes the kind of substantive answer we naturally expect.

The problem with the first is to explain what the conceptual confusion is supposed to be. This seems clear in the North Pole case: someone who would ask that question does not understand the concept of north.[11] The situation with the metaphysical question cannot parallel this, however, because it cannot be that someone who would ask why the world exists does not understand the concept of existence – the totality exists just as much as the parts do, and must, in fact, given that the parts do. So, it cannot be that the concept of existence loses traction when you get to the whole, in the way the concept of north might be thought to lose traction when you get to the Pole. The reason this might happen in the latter case would be because of a connection between the concepts of north and North Pole, such that to fully grasp the concept of north is to understand that nothing could be north of the Pole. But there is no such connection between the whole of reality and existence: the whole of reality not only *could* exist, but actually *does* – whereas if this conception of north was justified, then there *could not* be anything north of the North Pole, for conceptual reasons.

Perhaps the idea is that to ask why something exists is to ask for a cause, and that since the concept of the whole of reality is the concept of something which could not have a cause, anyone who grasps the significance of this form of question should realise that it is inapplicable to the case in hand. We are asking for the cause of something which, by its very nature, could not have one – because the cause would be part of it.

But that cannot be right either, because to ask why something exists is not necessarily to ask for a cause, despite the fact that a cause, or motive, is what we would normally expect. If it were, it would not make sense to seek alternative explanations once it is recognised – almost instantaneously – that a causal explanation will not work. If the question was simply a demand for a cause, then someone who subsequently toyed with the idea that a necessary being is responsible would be misunderstanding the question. But they are obviously not, because the question is simply a demand for an explanation – any explanation – of why reality exists; for relevant information that satisfies our curiosity.[12] To recognise that a causal explanation cannot provide the answer is to recognise that it is not a scientific question, and to think that this automatically renders the question insignificant is positivism at its crudest. It is not to explain why science cannot answer it, but rather to think that since it cannot, there must be some conceptual reason you are not allowed to ask, however desperate that reason might be. And indeed, desperate straits must have been reached for the scientistically inclined to be driven from empirical research to conceptual analysis.

So, what is it about reality that makes any explanation of its existence impossible? The answer, according to idealism, is that reality is transcendent. We start to see the force of this answer by remembering that to ask why reality exists is to ask why *ultimate* reality exists. If the ultimate reality were the physical universe, then it ought to be possible to explain why it exists, because we can explain individual physical existences, and the whole is the sum of its parts. But now suppose we entertain the idealist's suggestion that the physical universe is not the ultimate reality, but rather a way of understanding phenomena within conscious horizons. We suppose that consciousness encloses us within a phenomenal reality, the ultimate reality of which transcends it. This immediately explains why we should not be expecting a causal explanation: because the question is about ultimate reality, but causal explanations are only appropriate within phenomenal reality. They are appropriate there precisely *because* we are not trying to explain ultimate reality.

So, the reason we cannot answer the question causally, or with a motive (via an account of necessary being), or by any other conceivable means, is that these patterns of explanation are only applicable to the world we find within consciousness. When we ask why reality exists, however, we are asking about transcendent reality. And we have no reason to think that *any* notion of explanation we have has meaningful application to an ultimate, transcendent reality which we grasp in a manner so completely inadequate to its nature that, in metaphysical self-consciousness, we designate it 'transcendent'. So, knowing the paucity of what we can legitimately say about ultimate reality, we should not expect to be able to explain why it exists, as indeed we cannot. We have been provided with a reason to expect its existence to be a brute fact for us, namely that it transcends our experiential horizons. But the line of inquiry has been closed down through metaphysical insight, rather than a diktat. Rather than continue to live with our minds closed to an acknowledged mystery, idealism allows us to open them up to it and place it within the scope of reason and understanding.

(2) The Argument from Consciousness

Whatever the ultimate reality is, we must be part of it. If we were not, we would not exist. Things can be part of ultimate reality, however, without being interpreted as such. Hammers are part of ultimate reality, but our interpretation of them as hammers is not according to their ultimate nature. Hammers are found or crafted to fit our interpretation, and if the human race became a distant memory, any hammers we left behind might never again be interpreted as such. But the failure by those others to grasp their role within our form of life would not constitute a failure to grasp their ultimate nature. Some things have no ultimate nature, such as fictional characters; they are entirely dependent upon our interpretation, and hence do not exist. Hammers are not like that and neither are we.

Since we are part of ultimate reality, and are conscious, we must be conceiving ultimate reality as self-aware: as aware of itself, and in virtue of this awareness, potentially aware of other parts. It makes no sense for one part of reality to be an awareness of another without its also being an awareness of itself, because if the information the former bears about the latter is hidden from itself, then it is not aware of that information, and hence not aware of the other part. Awareness of another is self-awareness of information received from the other, so awareness of self is the basic experiential notion.

We conceive ourselves as many things other than as conscious – as animals, carbon-based, etc. – but all such interpretations, if construed as of the ultimate reality, presuppose that we must sometimes interpret it as self-aware. It must be self-aware, in part at least, in order for us to apply the 'animal' or 'carbon-based' interpretation to it, for if it were not, then reality would not self-consciously exist, and so these interpretations would never be applied. To offer any interpretation of ultimate reality whatsoever, then, requires us to interpret at least some parts of it as self-aware.

But though we are committed to thinking that at least some parts of the ultimate reality are self-aware, we are not thereby committed to thinking that our notion of 'self-awareness' is one which elucidates a feature of ultimate reality in an informative manner. We are only committed to thinking that *what* we think of as self-aware reality – conscious experience – is some part of the ultimate reality. In this way, we must succeed in accurately describing ultimate reality, or at least some of it. Accurate, that is, within the extremely broad range we are able to manage. And it must be broad indeed, because the experiences which inspire in us the thought of self-aware reality are conceived under the influence of objective thought. There are shapes to be discerned in visual experiences, and yet when you look at a square window, for example, there is no square shape to be found within a thing called an 'experience', since there are no such things in the objective world. We must conclude, then, that when we try to characterize the self-aware parts more precisely, by focusing on our experiences and describing them directly, the fact that we are obliged to employ objective thought shows that we fail. 'Self-aware' is the best we can manage. It focuses our attention on the right thing and nothing more.

But then, if our interpretation of the self-aware part of ultimate reality is accurate only to the extent that it tells us to think not of what you experience, but of the experience itself, then we have no reason to think that our other interpretations accurately characterize ultimate reality. We know that all of our interpretations of reality as something which is not self-aware are based on experience, and that our conception of experience tells us nothing about the ultimate reality, except that it exists. If a part of reality uses what it thinks of as its self-awareness to form an awareness of other parts, then the realisation that it has practically no grasp of its own ultimate nature removes any reason for expecting accuracy about the other parts. If all you know of the map, at the ultimate level, is that it exists, and that you cannot help thinking of it as a map, then at that level, you cannot expect to be forming a better conception of what it maps. So given that the self-aware parts have an absolutely minimal grasp of their ultimate nature, the other parts this grasp suggests are such that we cannot know their ultimate nature except in the same minimal way, albeit this time by implication from the self-awareness. And that is what the idealist believes. Another way of putting it is to say that ultimate reality is transcendent.

(3) The Argument from Concreteness

Numbers lack concrete reality – they are linguistic vessels without filling. If you want to say what the number four is, you can say it is twice two, half of eight, a third of twelve, and so on without limit. You can relate it to other numbers, but will never get to the thing itself, since there is none. With numbers, as Rorty once put it, there are 'relations all the way down, all the way up, and all the way out in every direction: you never reach something which is not just one more nexus of relations'.[13] But reality itself is not like that. It is not an abstract reality, but a concrete one, otherwise there would be nothing for us to relate, and no 'us' to do any relating.

Can we say what concrete reality is? Can we put it into words? We do not have to put numbers into words, because that is where they are already, they are a 'nexus of relations'. Given the words, the numbers follow. The same can be said of fictional characters, for if we take the works of Conan Doyle as canonical, everything there is to know about Sherlock Holmes is contained therein. We cannot sensibly ask what he had for breakfast on a certain morning if Conan Doyle did not mention it, for the Law of Excluded Middle (he either ate eggs or did not) does not apply to him, as it must with a real person, even if we will never know the answer because they are long dead. Verificationism is a sensible philosophy for a fictional world only.

The real world, however, is not created by what we say. So, you might start to wonder whether the idea of saying everything there is to say about ultimate reality, of 'capturing' it in words, makes much sense. As Plato realised long ago, it would make sense if reality consisted of ideas, for then our definition of goodness, for instance, would define the concrete reality itself. There would be nothing more to say about goodness apart from the relations it enters into, not because it is an abstract nexus, but because the idea those relations elicit in our minds instantiate its concrete essence.[14] But if reality is not like that, our words seem destined to leave it free and uncaptured, as they float off into the aether.

What we can do, incredibly well, is split concrete reality into parts which we relate to each other, thereby allowing us to predict and control our experiences. When we learn to relate parts to other parts, we think of ourselves as knowing what those parts are – trees, for example – and if we take it to the microscopic, and then sub-microscopic levels, we keep finding new relations. But we are still only relating concrete realities by treating them as empty linguistic vessels. As our explanatory models become more powerful and objectively all-encompassing, it becomes more obvious this is what we are doing. Contemporary physics is mathematical and mathematics is purely relational; the iconic $E=mc^2$ formula, for instance, tells you how to relate a number for energy, to a number for mass, to a number for speed. And quantum physics steadfastly resists anything but mathematical understanding – its formulae secure the predictions, and hence power, but even its leading exponents are baffled when they try to imagine a miniature reality they might characterise.[15]

So, our objective understanding of reality is relational. And our subjective understanding is too: we understand experiences by relating them to each other and to the objective conditions supposed to have caused them. When a wine connoisseur detects a hint of pencil-sharpenings in the flavour, this is illuminating only to the extent that it relates the flavour to a familiar smell. But then, if our descriptions of what we experience are relational, how are we to distinguish a description of reality from one of fiction? We need only look back to the history of science to find descriptions of an objective reality which is not our own, but if we were so minded, we could devise a description of another reality, replete with its own physical science, that is pure fantasy. We could do the same in metaphysics. Or rather than imagining alternative descriptions, imagine an alternative reality. Suppose we had the perfect objective and subjective description of our own reality: one which relates everything we could possibly experience to everything else. In an alternative reality that very same description would describe a fiction. What would be the difference?

The difference is that we are part of concrete reality and hence can see that the description fits the reality we belong to. Nothing in the description tells us this, however; for all it says, it could as well be a description of nothing at all. But we know there is a concrete reality, and if the description were of a fiction, there would be nothing within the world it described thinking this, even if the description said there was. So concrete reality must simply be *presupposed* when we make sense of it by relating its parts to each other, which is exactly how it seems when you think about experience: red is something, green is something else, and these somethings are presupposed, not elucidated, when we incorporate them into our explanatory web. We cannot 'say what they are' as we can say what a square is; once we have finished labelling and relating them, we just run out of words and leave the concrete reality standing where it was before.

But in metaphysics, we want to describe rather than simply presuppose the ultimate reality, and we want to at least register the fact that there is one. To do this, we cannot think of it as a phenomenon: as something that appears within our consciousness. For our understanding of phenomena is relational. It makes sense that it would be, because we need to understand things in terms of what they can do for us, and hence how they can feed into our projects. To describe the ultimate reality itself, however, we must think of it as something that transcends the phenomenal world which our consciousness sets up, and thus the relational understanding we have within it, where the fact of existence is simply presupposed. To understand reality as transcendent is still to understand it relationally, of course, but it is a relational understanding that gestures beyond itself, rather as chordal music can gesture towards atonality.

(4) The Argument from Time

There is nothing more concrete than the 'here' and 'now'. It is where objective thought acquires all the materials it makes relational sense of. And yet, when it tries to make sense of the 'here' and 'now' themselves, it finds only arbitrarily selected points of reference. Trivially, any part of the objective world can be treated as 'here' and 'now', and, if the part in question is supposed able to think, it would of course think of its place and time as 'here' and 'now'. But really, there is no 'here' and 'now' in the objective world, because the idea of an objective description is of one that negates perspective. Subjective thought anchors our thinking in the objective world, but only on the assumption that the sense of 'here' and 'now' which experience provides is trustworthy. In subjective thought's phenomenal mode, which is in hock to objective thought, if we suppose the world scattered with all the thoughts and feelings had by the humans and other animals capable of them, then again, each is only trivially 'here' and 'now' for their subject. Our anchor to the concrete, ultimate reality only comes to light through horizonal thinking. If we try to think of horizons scattered throughout the world, we simply misconceive them as phenomena. But we can reflect on our own horizon, or imagine ourselves into others one at a time, to find something ultimately real taking place 'now'.

We have conditional notions of both 'here' and 'now', but only an absolute notion of 'now'.[16] So, for instance, my current experience tells me that I am sitting at my desk. But in principle, I could have a dreaming or hallucinatory experience that falsely told me the same thing. So, this sense of 'here' is conditional: it tells me where I am located *if* my experience is veridical. We also have a conditional notion of 'now'. My experience tells me it is a Tuesday morning in 2018, but it could tell me the same in a dream or hallucination taking place at a different time. In addition to these conditional senses, however, we also have an absolute notion of 'now', which has no parallel in the case of spatial location. In this sense, my experience is 'now' whether or not it is veridical. For whether I am dreaming or having my brain manipulated, my experiences are still taking place 'now' – 'now', somewhere or another. If the experience is veridical, I can connect up this absolute 'now' with the conditional senses, to know when and where I am having the experiences. If it is not, I cannot, but it remains true that 'right now' someone is dreaming, or hallucinating, and that someone is me.

This absolute 'now' cannot be understood as a property of objective or subjective phenomena. As Aristotle showed, any attempt to make relational sense of it will result in it disappearing before your eyes. For the present must be durationless – anything with duration incorporates a past and future – which is just to say that it cannot be real at all. You cannot build up a continuum of time with durationless blocks. The British idealists Bradley and McTaggart showed the paradox which results from trying to think of the absolute 'now' as a property which privileges each step in the objective or subjective sequence of events, one at a time. This insight was later taken up by philosophers seeking a conception of time in line with relativity theory in physics, who came to the conclusion that no part of the four-dimensional space-time continuum can enjoy the privilege of absolute presentness, and hence that our lives are stretched out from beginning to end, rather than worked through progressively.

This is a natural consequence of following through on a perspectiveless conception of reality in metaphysics; of thinking how the phenomena must be conceived in order to be independent things, and forgetting the horizons they appear within. It makes our lives fade away to nothing. But since the absolute 'now' is as real as anything could be, what is it? Idealism holds that it is the transcendent reality, which we are trying to directly characterize so as to register the fact that there is a concrete reality we are part of. We see the force of this answer from the experiential examples which illustrate to us the distinction between the conditional and absolute notions of 'now'. The conditional relies on the experience being veridical to relate it to reality as understood in one way or another, but the absolute must characterize *something*. If the experience is illusory, the absolute 'now' points outside of experience and lands we know not where; all we know is that it must land on something which experiences the same absolute 'now'. If this is an objective body, or a subjective mind, then it will secure a conditional sense within the phenomenal reality of that body or mind. But it will have no place within it, and so will again point beyond, until recognition of transcendent reality is finally reached.

(5) The Argument from History

On current estimates, our species evolved about a quarter of a million years ago, on a four-and-a-half-billion-year-old planet, where life first appeared some time during its first billion years. Once we became civilized, about five thousand years ago, we rapidly colonized the planet, subduing all other animals and generally moulding the natural environment to suit our needs. During the last four hundred years, we mastered science and turbocharged our technological advance, with which we have been able to improve our quality of life, life-expectancy and dominance of the planet. That technology might give us eternal life is now considered a realistic prospect in some quarters, and we are trying to make lifelike machines to do our work and technological development for us. We are taking our first tentative steps towards leaving this planet for others, which we may eventually have to, since our population is rising at a historically unprecedented rate, due to our technological success.

Ever since our scientific revolution, which led to a succession of technological revolutions that transformed people's lives, the ancient philosophical idea that our science describes the ultimate nature of reality without remainder, namely materialism, has become more and more popular. It surged to an unprecedented level of popularity in the last century, when new technologies impacted our species like never before. It was the age when children expected to grow up to see flying cars; an age, in terms of the general kind of expectation, which remains with us. Since this ancient philosophy always ran into great difficulty trying to incorporate the experiences and thoughts which produce our science, its new followers questioned the very idea of experience, with some going all the way to a denial that experience even exists. Carried away by the spirit of the age, some people took this seriously, although hardly anyone was thinking about it at all.

On current estimates, our science describes only 4 per cent of the physical universe, which materialists think is the ultimate reality.[17] But we know enough to tentatively divide the rest of it between the two categories of 'dark matter' and 'dark energy': two different unknowns. Scientists are racing to discover more and their discoveries speed up our technological advance. Of course, our science may not correctly describe even 4%; maybe none of it. Nobody knows how much of our science will remain in a thousand years, should we continue that long. How much of our science from a thousand years ago remains I could not say. The apple still falls from the tree, but the best scientific description we had at that time of what makes it happen.... I doubt much of that remains. I would assume, as may simply be a product of the assuredness of our age, that much more of what we now have will survive. The scientific revolution of four hundred years ago seems like it will always be considered of great significance, but maybe not.

Now looked at from this historical perspective, how sensible does materialism sound? How sensible, given how much scientists know we cannot yet explain, and given what materialism is forced to say about the ordinary ideas we have always weaved our lives around and no doubt always will? What materialism says about these ideas, like experience or freedom, is always, sooner or later, either that they are illusory, or that although you would never have imagined it, physical science has been making sense of them all along. To me, that sounds like *wanting* our physics to be able to explain everything, despite the fact that the community of physical science is perfectly satisfied with the state of their art, as is everybody else, and is consequently quite unashamed of the fact that most of what it is uncontroversially supposed to be able to explain is currently beyond reach. Being able to estimate how much we cannot account for, and then divide it up a little, is an astonishing achievement, especially when you think what we knew a thousand years ago. But materialists want science to do so much more. Looked at from the outside, it is hard to imagine there being much reason behind this desire. And when you look on the inside, you find this suspicion confirmed beyond reasonable doubt.

There is another ancient tradition of philosophy, however, which is opposed to materialism. This is idealism, and in its various incarnations it was, until recently, more popular among philosophers than materialism. According to idealism, the ultimate reality transcends our experience and hence our descriptive powers, beyond the description made by the metaphysical claim itself. Does that not sound like a more realistic, sane and beneficial philosophy for our species at its current stage of development, if not forever?

Human beings might not have been the species that dominated this planet. Had it not been for that asteroid nearly sixty-six million years ago, perhaps some kind of reptile would have started on a trajectory similar to our own back in the Mesozoic. Maybe that species would have had a scientific revolution and ended up somewhere comparable to where we are now. Something like that might happen yet. Maybe we will use our technology to put an end to ourselves, following which a new species will start down our path. It might be a descendent of ants, or dolphins, or seagulls. In any case, imagine some other species developing rather as we did. They progress from their first civilizations to a scientific revolution within five thousand years, and thereafter things change faster and faster.

Now do you think that at some point in the development of their science, they would have to come to the same description of the physical universe as us? The same description of the 4 per cent; or perhaps the description of the whole thing we are fated to one day arrive at? Given that we would both be describing the same reality, you might think our differing scientific descriptions would have to be at least commensurable, such that what one was saying could be translated into the terminology of the other. But reality is multi-faceted in ways we know we have not yet grasped and may never do. Suppose this other species evolved from ants reliant upon ultra-violet vision. Suppose they have vastly different needs for their technology – perhaps entirely social needs and no personal ones. Would their science necessarily be like ours? If they could perform technological feats comparable to our own, such as leave the planet, would their science have to be either commensurable or wrong? The bare fact that they would be describing the same reality does not make this even slightly plausible. Human beings have all kinds of incommensurable vocabularies for the same reality – politics and chemistry, for instance – and we all have the same kinds of experiences, social practices and needs. But if this other species had a tradition of idealism, their understanding of ultimate reality might be so deliberately threadbare that it was indeed commensurable with our own. That we had been in possession of a philosophy like theirs might be one of the factors which made our extinction puzzling to them.

5

Technoparalysis

§1. A Positive Proposal

We live in the slipstream of ceaseless technological advance. We wait to see what new technological advances will come along and then strive to adapt to them after the event. We always have, but it has never been so obvious as now, given that everyone of at least middle age has witnessed how computers transformed our lives within a couple of decades; I belong to the first generation for which it was normal to grow up with computers. Perhaps there was pre-planning about how the internet would influence our societies, so as to mitigate the worst consequences envisaged. Perhaps some academics warned against going ahead. If so, it was to no avail. Nobody can be sure exactly what the new technologies will be, or what social problems they will generate, until the race to produce them is won; and then they rapidly transform when consumers and competitors get hold of them anyway. It is the race itself that needs to be tackled, now that the advance has become a problem. The reason our predicament has become so obvious of late is that we have all seen it getting faster, and are now in no doubt that left unchecked, it is will get faster still. 'Disruption' is the buzz-word among technology companies: disruption to our lives to make us dependent upon their new products, sweep away demand for established products (i.e. their competition), and thus finance superyachts – *and also*, according to some utopians, to force us out of our bad old habits so that the world becomes a better place. Whether or not ordinary people like the sound of the new wave of 'disruptions' now vaguely envisaged, there is currently no realistic prospect of disrupting *them*. So we just have to hope for the best. We have to hope we will number among those who will adapt and prosper, rather than fall by the wayside. We have to hope the social problems caused by the new addictions and powers will prove tractable. We have to hope that nobody comes up with something as bad, or worse, than nuclear and biological weapons. We live in a state of technoparalysis.

How do we break out of it? I think introducing more philosophy into our world would really help. This could be achieved by making philosophy a compulsory subject throughout primary and secondary school education – and not alongside religious studies, as it often is now, but as an independent subject. Philosophy needs space to reflect on all fields of human interest impartially. Starting early and keeping going will cultivate open-mindedness, and develop the ability and desire to engage in rational discussions about anything and everything. Back in the sixteenth century, in his classic essay 'Of the Education of Children', Montaigne had already seen the importance of trying to develop independence of mind through education. When a child is learning, 'if he embraces Xenophon and Plato's opinions by his own reasoning, they will no longer be theirs, they will be his', he wrote. And because we need to make ideas our own to really care about them, he recommended that 'the tutor make his charge pass everything through a sieve and lodge nothing in his head on mere authority and trust'.[1] This may not always be practical or desirable. But it would be in philosophy class. So it seems to me, as to many others, that there is a missing link in our children's education: an empowering one with universal application that provides the wider context we currently need. (The following endnote contains references to some of the extensive literature surrounding this educational agenda.[2])

In a world in which everyone had basic philosophical literacy, including all the scientists, technologists, consumers of technology and politicians, everyone would be used to thinking big: about the meaning of life, the nature of reality, the extent of human knowledge, the difference between right and wrong. If taught well, they would be used to thinking their opinions mattered as much anyone else's, so long as they were suitably informed; that reason and evidence were the only arbitrators. Those opinions, and the attitude behind them, could then filter into working practices, consumer habits and policies. Perhaps by that time we will already be living among robots, designer babies and cybernetically enhanced humans – if so, we always will be. But in such a world, we might decide to consolidate what we had at the time and learn to live with it – while we debated at leisure about whether, when, and in which fields, we should move forwards again.

We would not then live in a world of philosophers, any more than we currently live in a world of mathematicians simply on account of maths being universally taught – but philosophical issues, theories and standard argumentative moves would be familiar to us all. Edward O. Wilson, the biologist I mentioned in the introduction, has said that, 'The advances of science and technology will bring us the greatest moral dilemma since God stayed the hand of Abraham: how much to retrofit the human genotype.'[3] If so, we will be unequipped to deal with it if we remain on our current course and opinionated people like Wilson will decide for us. Most people are completely unaccustomed to thinking about general moral dilemmas, only the particular ones that happen to afflict them. We certainly do not want the decisions made through a combination of fear, where the money is smelled, and peer pressure. If ceaseless technological advance is going to generate real-world philosophical problems, by changing the basic conditions of human life, then widespread philosophical awareness will be needed to deal with it rationally. And this same awareness may help us to take collective control of the advance itself, so that the people it affects have more control over its pace and nature.

As things stand, about the most positive future most people seem able to imagine is one in which *after* the technologically precipitated apocalypse, the human spirit perseveres, as we heroically struggle to get ourselves back on our feet; post-apocalyptic drama is a thriving genre. It has become normal to suppose that we might need *that* kind of lesson, since from a prevalence of resigned pessimism, optimism has been invested in the thought that catastrophe might make all the difference. And if a significant contributing factor is that to imagine everything turning out fine is just too boring, then that once more indicates that we are not imagining our future very well at the moment.

There are deeply committed optimists, however, some calling themselves 'futurologists', who have other ideas – and some might even agree that the best way to predict what will happen to us is to steer things in the direction we want to go; at the individual level this could hardly be more obvious, since to secure the accuracy of my prediction that I am about to go to the shop, I need only get up and go to the shop. While we remain in the grip of scientism, however, the thought of *inevitability* will obstruct us.

§2. Inevitability and Desire

The internet pioneer and futurologist Kevin Kelly is a strong believer in inevitability – he has written a book called *The Inevitable: Understanding The 12 Technological Forces That Will Shape Our Future*. He combines this belief with some philosophy from Heraclitus:

> Because of technology everything we make is always in the process of becoming. Every kind of thing is becoming something else, while it churns from 'might' to 'is'. All is flux.[4]

The idea behind this is that 'the physics and mathematics that rule the dynamics of technology tend to favour certain behaviours' – they 'shape the general contours of technological forms' but 'do not govern specifics'. So a continual flux of technological advance is driven along by the physics in a direction beyond our control, but how exactly it actualizes is something we do have a say in. The internet was inevitable, he tells us, but not 'the specific kind of internet we chose to have'; telephones were inevitable, but not the iPhone. Kelly concludes that since we cannot fight the physics, we must learn to work with it: 'Massive tracking and total surveillance is here to stay. Ownership is shifting away. Virtual reality is becoming real. We can't stop artificial intelligences and robots from improving, creating new businesses, and taking our current jobs. It may be against our initial impulse, but we should embrace the perpetual remixing of these technologies.'[5]

Alternatively, you might think that it is human desires driving the technological advance, and that to the extent that the physics is playing a role, it is in determining how we are able to realise those desires; that the situation is effectively the reverse of what Kelly describes, except that the physics cannot have any good intentions when it works with our desires. The internet originated as military technology, funded in the hope that it would advantage the USA in the Cold War, and many of the key innovations which made it a worldwide phenomenon were pioneered by pornographers; you might say that *these* desires are at the root of everything positive we find in the internet.[6] A similar pattern of war then sex already seems to be emerging with the development of AI.[7] While we fail to reflect on the desires driving our technological advance, they become perfectly suited to play the role of the autonomous, unstoppable force that is Kelly's physics.

The British journalist, businessman and member of the House of Lords, Matt Ridley, agrees with Kelly about inevitability; for both, technology is an 'evolving organism' with its own independent desires. Ridley understands this in terms of a combination of Social Darwinism and laissez-faire economics: just as a 'recombination of genes as a result of sex' produces biological novelty, he says, so does the 'recombination of ideas as a result of trade' produce technological novelty – '"ideas having sex" explains why innovation has tended to happen in open societies indulging in enthusiastic free trade'.[8] This explains everything for Ridley, as indicated by the title of his book, *The Evolution of Everything* – but only everything *good*. When things go wrong it is the fault of meddling people. As he sums up his overall thesis: 'bad news is manmade, top-down, purposed stuff, imposed on history. Good news is accidental, unplanned, emergent stuff that gradually evolves'.[9] It may seem as if people have steered history, but they were just pawns in a game played by nobody: for every great scientist we celebrate, there were others waiting in the wings who would have made more or less the same discovery. And every other field of endeavour is like that too. Music, for example, 'changes under its own steam, with musicians carried along for the ride', such that 'Baroque begets classical begets romantic begets ragtime begets jazz begets blues begets rock begets pop'.[10] Ridley's idea that technological advance is like biological evolution is no exception: five authors (including himself and Kelly) published books promoting the idea between 2009 and 2011, he tells us.[11]

Ridley traces the origins of the inevitability idea to Lucretius, and ultimately Democritus – every chapter of *The Evolution of Everything* has an epigraph from Lucretius.[12] Thus, following the idea through to its contemporary incarnations in Dennett, Ridley thinks free will and consciousness are illusions.[13] He tells us that he wishes he had been taught Lucretius in school, as an antidote to the view that people and institutions shape the world, so that he could have seen past 'the illusion of design' to the 'emergent, unplanned, inexorable and beautiful process of change that lies beneath'. His experiences at school (he and his Eton friends dug tunnels to escape) often seem to feed directly into his views; especially his extreme right-wing agenda (prediction?) for education reform (evolution?). He is surprised that 'liberated, freethinking people' send their five-year-olds 'off to a sort of prison for the next twelve to sixteen years', where they are confined to 'cells called classrooms'. His solution is to let the market and technology take control: 'why not cut out the human almost entirely?' he asks, predicting that the, 'traditional university will surely be gone in fifty years, swept away by technology'.[14]

His views seem impervious to other significant experiences in his life, however. Ridley presided over the financial collapse of the Northern Rock bank in 2007, causing the first run on a British bank for 150 years, and requiring it to be renationalized to the cost of £27 billion. The Treasury Select Committee later concluded that his 'high-risk, reckless business strategy' was allowed to continue because the government overseeing agency had 'systematically failed in its regularity duty'. But Ridley disagrees: he thinks the financial crisis was caused by too much regulation – that it was 'a creationist, not an evolutionary phenomenon'.[15] He was just riding the physics, which will never let us down.

So was it inevitable that Caesar would cross the Rubicon? The true believer in this curious kind of inevitability will not feel obliged to answer 'yes', since it allows for human tinkering around the edges. They would try to keep their hypothesis unfalsifiable by saying, for instance, that the Republic was ripe for collapse at the time, and that if it had not been Caesar, it would have been someone else; just as the true believer in horoscopes will find some tall dark stranger they 'met' that day, even if they never left the house – perhaps they read about one. But it is not unfalsifiable; not within the scope of falsification the thesis sets up.

On 27 October 1962, Vasili Arkhipov was second in command on a Russian submarine that found itself caught up in an American naval blockade.[16] The Americans trapped and harried Arkhipov's submarine with training depth charges, in an effort to force it to the surface to identify itself; the Russians assumed these were proper depth charges, and the Americans did not know the submarine was armed with a five kiloton nuclear warhead. The Captain of the submarine thought the war had begun and was ready to fire, but the decision required two other officers to agree; one did, but Arkhipov could not be persuaded. There were various other near-misses during the Cuban Missile Crisis, and afterwards, but it is particularly clear in this case that, in all likelihood, the actions of one man prevented the supposedly benign evolutionary force from bringing itself to an end. Of course, complete disaster might yet have been avoided even if Arkhipov had agreed to fire, for maybe the warhead was a dud; or maybe both sides would have held back; or maybe some of us would have lived on through the nuclear winter. But there is no reason to think such an outcome was inevitable. Arkhipov did not have a bad influence and it is only bad influences that are allowed as an exception to the inevitability thesis. So we have a good reason to think it is false.[17]

There is something to it, however. People do not come up with ideas in a vacuum – the society they find themselves within has a big influence, and since that means lots of people simultaneously being influenced in similar ways, it is unsurprising that similar ideas often arise among different people. It was Thomas Carlyle, in his 1841 book, *On Heroes, Hero-Worship, and the Heroic in History*, who provided the ever-popular target of the 'Great Man Theory of History' – and Carlyle was perfectly aware of the currently influential alternative. He says:

> Show our critics a great man, a Luther for example, they begin to what they call "account" for him; not to worship him, but take the dimensions of him,—and bring him out to be a little kind of man! He was the "creature of the Time," they say; the Time called him forth, the Time did everything, he nothing—but what we the little critic could have done too! This seems to me but melancholy work.[18]

Although there is precious little to be said for Carlyle's elitist view that, 'The History of the World [...] was the biography of Great Men', there is at least this: some people find themselves in a position where their decisions can be of historical importance, and whenever there is innovation, there is often a leading figure who emerges, the first or among the first into that area, who exerts major influence. If we succumb to the thought of inevitability, then we will create a cultural environment in which the next generation of innovators and decision-makers do not so much innovate or decide, as simply implement the kind of future Kelly and Ridley can imagine; the one suggested by where we are now, which is disconcerting enough even if you do not reflect on how precarious our recent journey has been, the desires driving it, or on how it might develop past their imaginings.

If you do not like the look of where things seem to be heading, the thought of inevitability provides an excellent excuse for looking away. And if you do, the very same thought allows you to feel impregnable in your convictions. Kelly was right about the internet and wants to be right again, while Ridley wants unfettered capitalism. And what better way of persuading yourself you will get what you want than by means of the inevitability thought? But the twists and turns of our individual lives seem anything but inevitable, even with hindsight, and neither do developments on the world stage – look back to 1900 or 2000 and ask yourself how much of what was about to happen might then have seemed inevitable. We are not dealing with experience here, but with desires conjoined to a metaphysical theory, according to which the atoms which make up reality follow inevitable paths.

The inevitability tactic emerged as soon as materialism re-emerged in the Enlightenment. In *The System of Nature* (1770), Baron d'Holbach reasoned that since we are physical beings governed by the necessary laws of Nature, and since experience tells us that our nature is to act out of self-interest, then following through on that self-interest is bound to turn out for the best. We 'must take for the basis of morality the necessity of things', he said, and went on to have Nature Herself tell us that, 'it is I who punish, with an unerring hand, all the crimes of the earth; the wicked may escape the laws of man, but they never escape mine.'[19] Religions turn us against Nature's necessary laws, with Judaism being the worst.[20] Government does too, when it interferes with free-trade: 'The government should do nothing for the merchant except to leave him alone', he said.[21] Between Holbach and Ridley, the same kind of view was promoted by Herbert Spencer in the nineteenth century, who innovated to the extent that, having read Darwin, he was able to invoke Social Darwinism, an idea which remains very influential.

And yet, despite the dubiousness of the inevitability thesis, the direction of technological development does indeed have an air of inevitability to it at present, quite unlike other areas of endeavour, like the arts or politics. Why should that be? One reason, I think, is that because philosophy is now socially peripheral, the influence of the centuries-old tradition of materialist thinking about inevitability is going unnoticed. The materialist metaphysic itself plays a role in this, because it can so easily hide its philosophical status to become an embedded belief-system. Within the atmosphere this helps to create, the desires of those able to drive the technological advance – those with the money and power – are being left unchallenged. During the World Wars collective desire converged on winning, and technology leapt forwards as never before; the biplanes used at the beginning of the Second World War were obsolete six years later, when jet planes were making their debut. Then, after the wars, the desires of those who suffered most found some influence, and we saw developments such as votes for women and Britain's National Health Service. That technology is currently racing forward suggests convergence of desire.

What is the overarching desire that is driving technology forwards? My suggestion is that it is to *improve on the human race*. This is a distinct desire from wanting to *improve things for the human race*. The former would immediately explain why we are trying to make robots that are cleverer and stronger than we are, while simultaneously trying to physically alter human beings to make them cleverer, stronger, more attractive, more physically resilient and longer-lasting. Of course, if we can make the right kind of servants and remove the aspects of ourselves we do not like, then improving *on* the human race might improve things *for* the human race. But there are many other ways of trying to do the latter, such as political efforts aimed at conflict resolution and at reducing inequalities of wealth and opportunity.

Improving *on* the human race is just one idea for how to improve our lot – and it is the one suggested by materialism. For if you think of us as bits and pieces that can be taken apart and put back together again, it is the obvious solution. And of the two ideas that currently seem most promising for this, robots are better than genetic enhancement; the latter is merely a stopgap. For if you want the perfect hammer, there is no need to refine the hammer-like object you happen to have at your disposal, when you could start from scratch, using the best hammer-like objects as a model. And if there are no subjects of experience in the materialist's purely objective reality, as I think the position ultimately entails, then it makes no difference if the improvement on the human race is not actually *us*.

This particular path to betterment began with the scientific revolution of the seventeenth century. In the religious world-view it immediately challenged, we would simply find ourselves in an improved state after death – so long as we lived well. Then in the Enlightenment, building heaven on earth became an engineering project. As Isaiah Berlin sums it up, 'The programme seemed clear: one must scientifically find out what man consists of, and what he needs for his growth and satisfaction. When one had discovered what he is and what he requires, one will then ask where this last can be found; and then, by means of the appropriate inventions and discoveries, supply man's wants'.[22] The model for the programme was to be the heights of European civilization – these showed the way to the universal civilization to be built, and everything else that had happened was to be put out of mind. Voltaire made this attitude clear when he asked: 'If you have no more to tell us than that one barbarian succeeded another on the banks of the Oxus or the Ixartes, what use are you to the public?'[23] Voltaire, along with many other leaders of the Enlightenment, had the elitist attitude to history that Carlyle was later to popularise, which sometimes spilled over into racism.[24]

A practical, scientific approach to fulfilling the Enlightenment's dream of perfectibility was only to suggest itself in the nineteenth century, when the goal definitively switched to improving *on* the human race. The new approach was an almost immediate reaction to Darwin's theory of evolution, namely eugenics, which was both conceived and named by Darwin's cousin, Francis Galton, who amplified all the worst instincts of the Enlightenment. Galton proposed, among other things, that Africa be repopulated by the Chinese on account of their racial superiority to the indigenous population.[25] The idea of eugenics caught on like wildfire among intellectuals, until coming to an abrupt halt with the atrocities of the Nazi regime.[26] But with materialism as our backdrop, it is a hard idea to put down, especially now we have made so many advances in genetics. Richard Dawkins wonders whether, 'some 60 years after Hitler's death, we might at least venture to ask what the moral difference is between breeding for musical ability and forcing a child to take music lessons. Or why it is acceptable to train fast runners and high jumpers but not to breed them.'[27] He goes on to say that he thinks he could be persuaded that there is a crucial moral difference, but that we still ought to have the debate, because 60 years (70 now) is long enough after Hitler's death. Let us hope that scientists and technologists will wait until we can all have that debate – one at least as prominent as a US presidential election or the UK's referendum on leaving the European Union, but with greater sensitivity to truth and rational argument.

Whatever positive goals are used to sell genetic or cybernetic approaches to improving on the human race, the non-cutting-edge versions of these technologies – those not currently being used for military advantage by major powers – will allow people to buy advantages over other people, with the motivation for purchase being the prejudices we have tried so hard to overcome. This is diametrically opposed to the moral progress we have made in learning to value people as individuals, and consequently their achievements as those of individuals. It presupposes, as is typical of materialistic thinking, that our goals are objective things to get more and more of, but neglects the fact that the goals mean nothing unless we value them. Think of goals in that way and there is no point pursuing them unless you have bought the best modifications available, with drugs being the current frontrunner in many areas. But if you can just buy outcomes, they are no longer achievements – achievement will be confined to those producing the best 'mods', probably robots incapable of deriving any satisfaction from it – and the meaning we project onto our meaningless lives will fade away.

But modifying ourselves is only a stopgap in this traditional materialist agenda. As already noted, it makes more sense to build better beings from scratch, or just watch as the inevitable evolutionary force does it for us. The futurologist, transhumanist and singularitarian, Ray Kurzweil, predicts a 'Singularity', a 'future period during which the pace of technological change will be so rapid, its impact so deep, that human life will be irreversibly transformed' – and it is a surprisingly early 'future period' too (2045). When this 'fifth epoch' breaks upon us, we will merge with technology. Then with the sixth and final one, the 'universe wakes up', as 'intelligence, derived from its biological origins in human brains and its technological origins in human ingenuity, will begin to saturate the matter and energy in its midst. It will achieve this by reorganizing matter and energy to provide an optimal level of computation (...) to spread out from its origin on Earth.' This is the 'ultimate destiny of the Singularity and of the universe', and when it is reached, there will be a God; or, at least, something 'as close to God as [Kurzweil] can imagine'.[28] If you consistently combine materialism with the natural desire to make things better for human beings, then perhaps that makes good sense.

Reflecting on the dream of Kurzweil and other transhumanists, John Gray asks: 'Why should a post-human species have any value for humans?' The transhumanist must think 'a universe containing such a species would be a better place', he reasons. But since the universe has no point of view of its own to appreciate the improvement, and hence there is only ours and that of our successors to be considered, it is unclear what *our* interest might be in helping to enact our own replacement.[29] Gray is right about the confusion, but overlooks the reason for it. The reason is that materialism makes it very difficult to believe that there are any points of view. Since the transhumanist must employ one when their misanthropy finds satisfaction in the thought of our demise, they end up thinking the universe would be better for us without us. Gray goes on, however, to make the definite practical objection to dreams of post-human utopias, and with great style too: by pointing out that there is no chance of us evolving into God in the world we actually live in, only 'a warring pantheon of gods. Anyone who wants a glimpse of what a post-human future might be like should read Homer'.[30]

Metaphysical views go so deep that their effects can be felt anywhere, without their origins being noticed. If those views are false then we are collectively out of touch with reality, and the social view they have inspired may be detrimental to us; we all know the detrimental effects that can occur when *individuals* get out of touch with reality. The desire to improve on the human race has only achieved a fringe self-consciousness, among transhumanists, so it is not an overarching desire akin to everyone working together to win a war. But it is a desire which makes good sense of the current direction of technological advance.

The way we address this problem cannot be top-down. If somebody of the stature of Stephen Hawking can be convinced, as he was, that the development of AI spells doom for us all, but without it making any difference, then the individual views of influential people seem very unlikely to ever have much of an effect. Even if the president of a major world power were to make the problem of ceaseless technological advance their top priority, this would only disadvantage their country in the technological race. The change can only be bottom-up, spreading through new generations, and on the basis of individuals thinking for themselves about our relationship with technology. Scientism breeds thoughtlessness and apathy, but philosophy encourages thought and concern.

Materialists who recognise the problem of ceaseless technological advance have different ideas for how to solve it, however. The boldest proposal of this kind is made in Ingmar Persson and Julian Savulescu's book, *Unfit for the Future: The Need for Moral Enhancement*. They doubt we will be able to indefinitely keep a grip on the powers science and technology are unlocking at an accelerating rate, and think we are consequently at ever-increasing risk of 'Ultimate Harm', as they put it. They say that even 'present technological know-how makes it possible for small groups, or even single individuals, to kill millions of us'.[31] They argue that we need to think differently about scientific and technological advance, and propose a new word for 'moral wisdom as regards the pursuit of scientific research and its practical applications': 'science-sophy'.[32] This latter idea is one I broadly agree with, and resembles Nicholas Maxwell's call for a new, wisdom-centred approach to inquiry, which is something he has spent decades arguing tirelessly for.[33] And yet Persson and Savulescu's main proposal, as the subtitle of their book reveals, is to 'mod' the human race. They think 'some children should be subjected to bio-enhancement, just as they are now subjected to traditional moral education'. Which children, I wonder? I could not isolate the answer in the text, although they say it would be 'the majority of people'.[34]

Dawkins was merely unsure whether there is a crucial difference between 'breeding for musical ability and forcing a child to take music lessons', but Persson and Savulescu lack equivalent doubts. They think that being subjected to either bio-enhancement or 'traditional moral education' is morally equivalent.[35] But the difference, surely, is that you can reject what you were taught but you cannot reject what is implanted into you.[36] I do not doubt that there is plenty of bad, dictatorial teaching still around, and even indoctrination. But things are changing, and moral equivalence with that kind of acknowledged bad practice is hardly a selling point for bio-enhancement, even if it could be attained. And it cannot for the simple reason that if bad teaching cannot be resisted at the time – daydreaming is a traditional technique – then it will still end, leaving the person free to go their own way. Modifying somebody to make them think a certain way is forcing them to think what you think. It is a violation of freedom of thought, the one freedom even the most extreme totalitarian regimes of our history have found themselves unable to curtail. Any enhancement considered moral by people prepared to enforce a programme of this kind seems unlikely to be moral at all.

If that is the only way to solve the problem of ceaseless technological advance, then I suggest we carry on hoping for the best. But it is evidently not. Persson and Savulescu have seen the problem and want it addressed. If encouraged to think these matters over, as we encourage people to think about maths, history and science, might not large numbers of people recognise the problem and want it addressed too? The most appropriate class for that to happen in, I suggest, would be philosophy class.

§3. The Need for Greater Philosophical Awareness

The problem with relying on technological advances to address the problem of ceaseless technological advance – with trying to use technology to make its ceaselessness safe, such as with moral enhancement – is that the same advances will generate their own dangers. If you can morally enhance people, you can immorally enhance them too, and the groups you are worried about will take this option; which they will not characterize in this manner, of course. The problem has arisen because the technological advance is not guided by collective rationality, and a contributing factor is that the materialist thinking embedded in our history has made it come to seem like an inevitable force outside of our control, while simultaneously helping to make the very notion of philosophical reflection invisible. The problem itself is far too big to be solved by any one proposal, but if I am right about the relevance of philosophy, then matters would be helped if it was taught much more widely than it currently is. Or to put it rather more idealistically, as seems fitting, I think the next stage of our development must be philosophical. And that means starting young: opening a new generation of minds to the biggest thoughts before our unphilosophical world has a chance to close them up again, as the older generation steps back to watch the sparks of an expanded collective rationality fly up.

Of course, the spread of philosophical literacy might make little or no difference to apathy about technological advance. But it might make a big difference, if it made people more responsive to arguments, readier to ask themselves: 'has this person got a point?' It would create more familiarity with the widest possible contexts in which we can think about our lives. And it would squarely address the problem, often remarked upon in the philosophy of technology, of the difficulty of isolating the point at which decisions are made within the genesis of new technologies, so that ethical considerations can be introduced. Everyone, from the teams of engineers and scientists, to the management structures, to those providing the financial investment, would be more familiar with the issues than they are now, and more able to rationally engage with them. Or, as I would prefer to put it, and already did in the last chapter, we might come to accord the very highest value to *interested* inquiry in science and technology, and become used to engaging in rational debate about what those interests should be.

It is often rational to deliberate about what facts you are comfortable with uncovering, before endeavouring to make a discovery. Given knowledge of the kinds of things you may find out, which you need at the beginning of any inquiry, you may decide you would be better off not knowing. In a world in which endless knowledge is possible, such deliberation is very important, not only because you need to think about how best to direct your efforts, but because if you do not so deliberate, you may cause great harm. Imagine a sealed envelope which, for whatever reason, contains the date of your death. Since opening it would probably ruin your life – unless you were in special circumstances or had a perverse psychology – then you would be a fool to do so. If you succumbed to curiosity, this would reveal either a failure to think through the consequences, or an inability to act on the basis of your deliberations. It would be irrational. And it could be immoral too, if there were foreseeably negative consequences for the people around you.

We already take this for granted in certain sensitive areas, which is why scientific research into whether race and IQ are connected has been a subject of continual controversy.$_{37}$ This project is rightly starved of funding, and any scientist who was neither allied with, nor trying to counter, a racist agenda, would feel very uneasy about getting involved. But we have come to think that investigating how to control reality at the subatomic level, how to artificially reproduce the cognitive functioning of free and conscious people, and how to predict, alter and control that functioning within people themselves, are not similarly sensitive areas.

Reflection on the direction of our technological advance is not something many people think about, even when buying into new technologies which, through a foreseeable further development, might take away their jobs. This is encouraged by our largely un-reflected materialist backdrop, which obscures and trivializes the possibility of philosophical reflection, while providing a metaphysical incentive to scientists to discover everything they can – to leave no truths covered by ignorance. And yet there is no end to the facts we could uncover, most of which we would find utterly boring; the ongoing location of some rock on a distant planet, for instance, or just at the bottom of one of our own oceans.$_{38}$ Obviously, the process must be directed. This directing of aims and desires needs to be governed by a more collective and consensual rationality. Greater philosophical awareness would surely help.

And yet disinterested inquiry of all kinds might be thought to embody one of most admirable and distinctively human of our traits: the spirit of adventure. This spirit is encapsulated by the response the ill-fated mountaineer George Mallory is supposed to have given when asked why he wanted to climb Mount Everest: 'Because it's there'. Adventurousness of this kind has been a great benefit to us collectively, and evolutionary psychologists even have a story to tell about why it propagates successful genes; the adventurous part of the brain has been isolated, apparently – although I find all such claims highly dubious until provided with a careful philosophical interpretation, of which many are always possible.$_{39}$

But there is no reason to think that more philosophical reflection and debate around research that might foreseeably feed the technological advance in ways that affect people generally would have any tendency to give us qualms about disinterested inquiry in other areas, or indeed just doing things for the hell of it when there are no foreseeable risks, or when the risk is only to ourselves and to others who agree to it.[40] The proposal is only that we stop valorising disinterested research in sensitive areas, that research in such areas must be mandated by public debate, and that greater philosophical awareness will make it easier to recognise which areas these are and help to generate more interest and involvement in the debates surrounding them.

It might be objected that this cannot be achieved except by banning certain areas of science, or, if the scientists themselves are supposed to self-govern, then their only realistic option would be to give up. For nobody can predict what *any* scientific investigation will discover until it discovers it, it might be said. That is why it is called *discovery*; facts that are covered over cannot warn us of their potential consequences. Many of the greatest scientific discoveries have been made by chance, with penicillin, X-rays and DNA being some prime examples. It has been shown, by psychologists Kevin Dunbar and Jonathan Fugelsang, that blind chance plays a crucial role in ordinary scientific research, with a major one-year study showing that over half of the findings made by the scientists under investigation were unexpected.[41] They found that 'rather than being a rare event, the unexpected finding was a regular event', such that 'scientists spend a great deal of time reasoning about unexpected findings'; that they designed experiments to produce a variety of them, and devised strategies for taking best advantage of them.[42] With chance this integral to the scientific method, then, perhaps it makes little or no difference whether scientists, or anyone else, thinks about the possible consequences of their work, since they have no way of knowing what they will find.

But this is not the strong objection it first seems. For Dunbar and Fugelsang also observe that, 'Human beings, in general, and scientists, in particular, appear to have a propensity to attend to the unexpected, have definitive strategies for sifting through these findings, and focus on the potentially interesting'.[43] And why are some results more interesting? Because there is a chance they might lead somewhere, possibly in line with the aims of the research project, but possibly somewhere else requiring a new funding application. Thus may begin a voyage of discovery, in which there will be plenty of time for concerns about potential technological applications to arise. The concerns could be brought into the public arena if the potential was serious enough. The path between the surprising discovery, further investigation, and technological application is a long one, and we could stop racing down it.

Very few scientists, proportionally, are involved with research in which this kind of issue could arise, and in the cases of those who are, the burden of moral responsibility must be shared with business leaders, engineers, politicians and the general public. The idea that ethics panels operating behind closed doors are an adequate safeguard for our collective future, however, is amply refuted by where we are now. Where we are now, it is hard to imagine anyone implementing Nick Bostrum's suggestion that 'a preemptive strike on a sovereign nation' could be justified simply on the grounds of 'the mere decision to go forward with development of the hazardous technology in the absence of sufficient regulation', since this can put 'the rest of the rest of the world at an even greater risk than would, say, firing off several nuclear missiles in random directions.'$_{44}$ Whether this could ever be justified I do not know, but I certainly do not think it is up to philosophers like Bostrum or myself, nor the hypothetic scientists developing the hazardous technology, to make such decisions.

This should be clear from the seriousness of the conflict of interests at the heart of the issue. Any restraint to our technological advance, no matter how it is done, may lead to unnecessary suffering; suffering, that is, which is unnecessary in relation to what we might have been able to do with the products of a less restrained technological advance. Continuing down our current path with AI research, for example, may allow us to save the lives of children who would otherwise waste away from diseases we cannot presently treat. It may save the whole human race, by allowing us to stop a worldwide pandemic, or reverse climate change. On the other hand, we may use it to ruin the best things about the life we have now, or bring about our own demise, as so many worry that we will.

Perhaps most people, after thinking it through, would prefer the excitement of riding the advance to see where it takes us. Perhaps most would expect the benefits to always outweigh the risks. Perhaps most would think that we have already caused so many problems that only a relatively or completely unrestrained advance can save us. But if most people do not think about it at all, as at present, then the advance will not be directed by collective rationality and consent. The mechanisms for informing public debates on these matters are already building, but one crucial element is missing: the one which gets people interested and which cannot amount to just telling them.

The current deadlock, technoparalysis, is one in which optimism reigns over a kingdom of resigned pessimism, indifference and ignorance, and this provides just as little chance of legitimate concerns being actioned, as of the optimism being justified to legitimise the reign. The deadlock is one of reason, which accelerates the ceaseless technological advance. It is an exclusive focus on the titans and a forgetting of the gods. Education is a sensible approach to addressing a deadlock of reason. If you think of philosophy as a disciplined and universal form of reflection which can take into account any kind of consideration whatsoever, so long as it can pass the gates of reason, then I think filling this gap in our educational agenda might be especially helpful in addressing this deadlock. It is the only reflection of this kind we possess; people respond to it when given half a chance, especially children; the response is proactive; and philosophy's unique universality is demonstrated by the fact that the academic discipline contains philosophy of science, philosophy of religion, philosophy of art, environmental philosophy … you name it.[45] 'Since it is philosophy that teaches us to live, and since there is a lesson in it for childhood as well as for the other ages, why is it not imparted to children?'[46] If 'teaches us to live' is interpreted broadly and collectively, then our long-overdue answer to Montaigne's question should be: 'we will'.

§4. Better Angels and *Homo rapiens*

For Steven Pinker, scientific progress and Enlightenment rationality has brought out the 'better angels of our nature', as he puts it in the title of his popular bestseller, and if that process of pacification is to be continued we need more of the same: we need 'enlightenment now', as the title of the sequel declares. For Pinker, John Gray is a prime example of an elitist intellectual who blinds himself to the overwhelming empirical data demonstrating the progress we have made towards a happier and less violent world, through a misplaced nostalgia for a past of ignorance and barbarism, and lack of respect for what global capitalism and science has done for the billions of ordinary people who now reap the benefits. For Gray, on the other hand, we are *Homo rapiens*, an incurably ravenous and destructive species which has raped the natural environment for its own ends, and with a population now so out of control that mother Gaia may soon shrug off the pestilence; she may trample us like the 'straw dogs' used in ancient Chinese rituals, first revered, then discarded, as he announces in the title of his own popular bestseller. For Gray, Pinker is a prime example of a conceited and historically ignorant neoliberal humanist, who has unwittingly taken to heart the Christian illusion of continuous progress, and thereby blinded himself to the meaningless cycles of fragile progress and consequent decline to be found throughout our history, as well as the much darker picture of life that real science, and particularly Darwinian evolution, reveals.

For the better angels and *Homo rapiens*, there is no problem of ceaseless technological advance. The better angels solve their problems with the advance: technology is 'not the reason our species must someday face the Grim Reaper', but rather 'our best hope for cheating death, at least for a while'.[47] *Homo rapiens*, by contrast, 'use their growing knowledge in the service of their most urgent needs – however conflicting, or ultimately self-destructive, these may prove to be'.[48] Technology is not something they can control, so there can hardly be a problem of how they should control it, since it is rather 'an event that has befallen the world' which 'leaves only one problem unsolved: the frailty of human nature' - and this problem is 'insoluble'.[49] So on the one hand, we have the familiar 'solve problems created by the last wave of technology with the next' thought, and on the other, we have a novel version of the inevitability idea, where technology is no longer an intrinsically benign external force, but rather an external force which can be used for good or ill, with our own immutable natures making it inevitable we will use it for ill.

Pinker is a materialist and so provides a typically materialist response; although he is refreshingly free of anti-philosophy and does not willingly fall in with the inevitability crowd either.[50] Gray rejects materialism for entirely sensible reasons, but advocates no alternative metaphysic. He is a sceptic, in this regard and many others, who does occasionally espouse some very ill-considered anti-philosophy.[51] His response to this issue has little to do with scepticism, however, since it rests on a commitment to human beings having an immutable nature. His inspiration, the idealist Schopenhauer, held a similar commitment, but Gray's seems to be essentially materialist.

To try to adjudicate the very public and polarized Pinker/Gray dispute, let us begin with Pinker, whose headline project of showing that the world is becoming a less violent place provides his premise to argue that technological advance does not threaten our survival, and that although things could go disastrously wrong, as he concedes, the chances are grossly exaggerated. 'The storyline that climaxes in harsh payback for technological hubris', he says, 'is an archetype of Western fiction, including Promethean fire, Pandora's box, Icarus's flight, Faust's bargain, the Sorcerer's Apprentice, Frankenstein's monster ...' – it is a *storyline*, not something to be taken seriously, and perhaps the reason people like me do take it seriously, he speculates, is that the fantasy of living at the End of Days adds poignancy to life.[52] Alternatively, you might think the stories convey a moral which has very recently come of age. Suppose the likelihood of destructive acts continually decreases, as Pinker thinks will happen, while the destructive potential of those acts continually increases, as it plainly has and is continuing to do so. Until the likelihood reaches zero, then there is something to worry about, and when the capacity reaches the point where it can end the whole process, as it already has, something very serious to worry about. Suppose you receive minor electric shocks frequently in the morning, by midday they are less frequent but becoming rather painful, and by the evening they seem to have stopped, but you know that if one happens now it will kill you – you have a lot more to worry about than before, surely. If you were thinking about it, this long evening peace would seem rather eerie.

Now Pinker thinks existential threats are declining because there has been a 'change within the mainstream of the developed world (and increasingly, the rest of the world) in the shared cognitive categorization of war'.[53] Essentially, we now stigmatize it, and that explains the Long Peace (1945–91) and what he calls the 'New Peace' (1991 onwards). It is typically assumed that the threat of nuclear annihilation kept the Cold War cold ('no serious military historian doubts this', says Gray), but Pinker disagrees. He thinks it was our new, more peaceful attitude, such that a nuclear strike would only be contemplated in the case of an existential threat. He cites as evidence the fact that non-nuclear Argentina invaded the Falkland Islands without fear of nuclear retaliation from Britain.[54] I cannot see how this could evidence a more peaceful attitude unless the aim of all wars prior to the Long Peace was to storm the enemy's capital and put everyone to the sword, which is obviously untrue. But nevertheless, just suppose Pinker is right. Are we now so peaceful that a confrontation in which an existential threat is perceived between nuclear powers could never occur again? Of course not, as Pinker accepts, but he insists that the threat is exaggerated and that technological advance is not to blame.

Pinker's favoured approach to concern about existential threat is particularistic, namely to work through all the main areas of concern at present – nuclear war, terrorism with biological weapons, AI getting out of our control, etc. – to show that anyone who knows what they are talking about knows the risks are negligible. His reassurances are hardly sophisticated. Most terrorists are 'bumbling schlemiels', for example, and if we developed a problem with AI we could just hit the off-switch – 'One can always imagine a Doomsday Computer that is malevolent, universally empowered, always on, and tamperproof', but the 'way to deal with this threat is straightforward: don't build one.' As far as more general considerations go, Pinker cites Kelly in support of the idea that, 'The more powerful technologies become, the more socially embedded they become', which means more societal influence, more safeguards, and hence less chance of maverick usage. So a team of scientists trying to take down the internet or exterminate humanity with a bio-weapon 'probably still couldn't do it', as he quotes Kelly saying, because they would be up against 'hundreds of thousands of man-years of effort' in the former case, and 'millions of years of evolutionary effort' in the latter.[55]

Pinker finds Kelly's reasoning 'abstract', but cites it nonetheless as a counterpoise to the similarly abstract (thus: bad?) view that the advance means 'humanity is screwed'.[56] Pinker prefers particulars and empirical data. But there are two big problems with listening to materialists on particulars. The first is knowing who to listen to. According to Dennett, for instance, the 'internet is fragile', planning for what to do if it crashes has been insufficient, and yet it 'has become the planet's nervous system, and without it, we are all toast'.[57] Persson and Savulescu, as I mentioned before, tell us that one person can now 'kill millions of us'. But according to Pinker and Kelly, the internet is incredibly robust, and even a team of dedicated scientists could probably not kill millions. Disputed factual states of affairs are an absurdly flimsy basis for addressing very general lines of reasoning, but this is what scientistic culture encourages.

The second big problem with listening to materialists on particulars is that it promotes myopia. Dennett, in the article just cited, provides a good example.

He was commenting on the threat of AI exceeding our intelligence and subsequently turning on us, and his first instinct was to say that it is 'imprudent' to even think about this, since it deflects attention from the very real threat we currently face of the internet being brought down. 'Stick with what you know', says one materialist, 'we know nothing of the sort' says another... and the line of reasoning which began the dispute quietly leaves the scene.

You cannot reason without facts, of course, if only stipulated facts. But facts will not do our reasoning for us. Which facts are pertinent depends on the level of generality of the reasoning. The facts that our destructive capacities grew to unprecedented levels within our very recent history and continue to grow, that we have had some near misses, and that there are serious international tensions in the world today, are of the appropriate level of generality to establish that we have a problem of ceaseless technological advance – and they are all facts which are beyond dispute. Pinker's argument that we have become more peaceful over the course of history is certainly of the right level of generality to connect with this problem, but his attempts to downplay the particular present threats are not. The way the pacification argument might connect with the problem would be by giving us reason to believe it will naturally resolve itself, so long as we are not unlucky in the interim – we will get more powerful and more peaceful in tandem, so the powers will probably never be used for ill. However, Pinker's explanation of why things sometimes do go disastrously wrong completely undermines any such argument.

It comes straight from Holbach's inevitability tradition, which Pinker had the good sense to otherwise distance himself from. It is the 'Great Man' theory of disaster in history. Thus the Second World War was the fault of one man, according to Pinker's first book; without Hitler it would have probably never happened.[58] In the sequel, he changes his mind about the culprit: it was Nietzsche – a claim which gave Gray a field day when reviewing the book.[59] But look away from the implausibility once more and just suppose this is all true, so that we can stick with the line of reasoning. In that case, the pacification process will fail to address the problem of ceaseless technological advance, because sooner or later, another Great Man will come along to wreak havoc with the best tech of his time. In the preface to *Enlightenment Now*, Pinker begins by expressing his strong disapproval of Donald Trump, the most powerful man in the world today. Great Men whose influence Pinker does not like are clearly not a thing of the past, then. Francis Fukuyama retracted his famous claim that history has ended when he started to reflect on the ceaseless technological advance, and given what Pinker's closely related views actually amount to, the core Enlightenment value of reason which he champions would seem to dictate that he do the same.[60]

The problem with Pinker's kind of optimism is that it could only be conclusively proved wrong when it no longer mattered. And that is because it is too conclusive an optimism – its conclusiveness is not scaled to the acknowledged possibility, because the acknowledged possibility is complete catastrophe. In the world we live in, I think you would have to know something very special indeed – hitherto unknown and as conclusive as could be expected in such a matter – to be able to justifiably present as optimistic a case as Pinker does on a matter of this ultimate scale. Pinker knows nothing of the kind. He is not appreciating the gravity. To do so is not to make pessimism and resignation remotely inevitable, but this is the concern that seems to spur him on.

Turning now to Gray, the first thing to note is that the headlines he uses to publicise his views are misleading. For despite his continual talk of the 'illusion' of progress, he does not really deny that we have made progress; once the qualifications are made, little remains of the 'illusion' and 'non-progress' headlines. Neither does his talk of *Homo rapiens* reflect any great misanthropy, I do not think; not much. Rather, the serious intent behind the shock value of these devices is to disrupt the kind of discourse, typified by Pinker, according to which science, technology and global capitalism are driving inexorable human progress on all fronts.

Progress in science and technology is real enough, for Gray, and so is progress in politics and morality – but they are very different and not connected. The former is a knowledge-base that builds incrementally and never gets lost, but the latter is a practical skill set, passed down from generation to generation, that is very easily lost. He illustrates this latter point by discussing renewed enthusiasm for judicial torture among some American liberals – renewed interest in eugenics is another prime example. He thinks that linking these two very different kinds of progress is a dangerous idea stemming from the Enlightenment, and was the cause of the mass carnage caused by the Maoist, Soviet and Nazi regimes in the twentieth century, all of which saw themselves as progressive regimes using science and technology to alter and perfect the human condition. And the ultimate cause of this dangerous folly, he thinks, is trying to tether the Christian belief in incremental moral progress towards a state of perfection, with the essentially amoral progress achieved in science and technology.[61]

Blaming all our woes on religion is a tiresome genre which Nietzsche exhausted long ago, and with more panache than is ever managed nowadays, as Gray might agree. He is particularly scathing of the 'new atheists' who say religion should not be tolerated since it consists in a set of propositions that science has demonstrated to be false – quite unlike the similarly intolerant attacks on religion which were perpetrated by 'secular faiths' of the twentieth century, Gray notes that this is 'a media phenomenon ... best appreciated as a type of entertainment'.[62] Against this kind of discourse, Gray emphasises the naturalness of religion:

It is a strange sort of naturalism that singles out religion to be purged from human life. Few things are more natural for humans than religion. To be sure, religion has brought much suffering. So has love and the pursuit of knowledge. Like them, religion is part of being human.[63]

The kind of atheist who passionately rejects religion has repressed this natural instinct, Gray thinks, then invested its irrational passion into science and technology. And in this way, the repression has produced, and will continue to produce, grotesque results.

But despite all these distancing manoeuvres, the view that religion is ultimately to blame for trust in the advance as a force for unmitigated moral progress is right there at the heart of Gray's position. Yet the Communist regimes were overtly atheistic and materialistic, the Nazi regime's relation to religion was ambivalent to say the least (there was great enthusiasm for paganism among some of it leaders, which Gray considers a much wiser form of the religious impulse), and the contemporary race for technological advance has no overt religious underpinning – quite the opposite. As for the naturalness of religion, I would say only that asking philosophical questions is a natural, rational instinct, and that if you try to suppress it through a desire for closure, then religion, scientism, or some other reason not to ask anymore is a natural enough consequence; one which can breed intolerance both for those who continue to inquire, and for those who opt for a different kind of closure.

Calling us all '*Homo rapiens*' is rather distasteful, but Gray is trying to make a serious point. He is trying to 'present a view of things in which humans are not central', as he says at the outset of *Straw Dogs*, and if you look at us as just one species among others, that is indeed how we currently look.[64] Our population and living standards are rising, and hence our appetite for natural resources is becoming ever more ravenous, with the consequences now clearly showing; Greenland, one of the few remaining wildernesses, currently has various powers vying for the chance to open it up to mass development. Pinker is at his least convincing when he tries to argue that even in the case of the environment we are making good progress. Japan has recently recommenced commercial whaling. The Brazilian rainforest is being gradually cleared. Almost all of the most amazing and iconic animals which children around the world learn to name as soon as they learn to talk have seen their numbers radically depleted during the last century, with many on the verge of extinction; that limited efforts have finally been made to do something about it, and some extinctions have been averted, is supposed to tip the balance in favour of general progress, according to Pinker.[65] To this extent, Gray has a point, and perhaps the provocative way he makes it has some purpose.

But this is not all that Gray means. He thinks our desires for dominance and destruction are just as natural as our desires for love and peace; that all these contradictory instincts characterize the human animal. He thinks that Darwin's theory of evolution, which makes no mention of moral progress – or progress of any kind – revealed a great truth which belies the religiously inspired faith in progress of neoliberals. It is in the nature of *Homo rapiens* to grab everything in its path in pursuit of its immediate interests, and we cannot fight our own natures, only repress them and thereby make matters worse. As far as positive suggestions go, in fact, the only advice Gray really offers is that we recognise the repressed religious instinct that produces a faith in progress, and that this instinct can engender deluded and dangerous large-scale plans for social change.

And yet he accepts that we *can* make moral and political progress. As Pinker documents in meticulous detail, we have been making lots of this for a sustained period of time now. Abject poverty is in serious decline and hitting children, sexism, racism, disability prejudice and homophobia are all now stigmatised, as they were not within living memory. Not everywhere, of course, but these skill sets, as Gray thinks of them, are entrenched in large parts of the world. If some crisis of resources made our living conditions more precarious, Gray would not start killing and torturing. He might take and sanction more desperate and selfish measures, as might any of us, but it would surely take many generations for our skill sets to be entirely lost – and since our way of life is so well-documented now, we could get them back. Is a major collapse of that kind bound to happen because of the cyclical nature of history? That sounds like superstition to me. The Roman Empire collapsed, which is the clearest example history provides. But in the Renaissance we recovered the insights and advances of the ancient world.

Are our inner, barbarous natures supposed to prevent our progressing indefinitely? Some existentialists believe that we have no inner natures and are entirely self-creating. In-between those extremes, you find a real world in which nurture has left nature far behind, and in which people increasingly satisfy their fascination with violence through television and video games. Reason, history and society make our evolutionary origins in a struggle for survival increasingly irrelevant, and although a devotee might try to tie everything we do to evolution, this is not a description of ourselves we are obliged to pursue. What seems to make Gray think otherwise is his commitment to an inner nature of the human animal, immutable by anything except impersonal evolutionary pressures. That is a very strong commitment for a sceptic, and one which a materialist might consider obligatory, but Gray should not.

Gray may be right that new technologies will be used to perpetrate future atrocities, just as the old ones were. But I see no reason to think this, or anything else, is inevitable. The idea that the technological advance is an unstoppable force beyond our control is just the same old materialist thinking you find in Kelly, Ridley, and the rest of the Holbach tradition; Gray simply removes the optimism. But the advance is really just something created by human beings through choice. Gray is sceptical about choice, whether individual or collective:

> Our lives are more like fragmentary dreams than the enactments of conscious selves. We control very little of what we most care about; many of our most fateful decisions are made unbeknownst to ourselves. Yet we insist that *mankind* can achieve what *we* cannot: conscious mastery of its existence.[66]

Life can be messy, but you play the hands you are dealt, try to stay on-track, and sometimes manage it. At a collective level, things get even messier, but mankind has chosen to respect moral codes, to live by the rule of law, and to stigmatise slavery and war. We could choose to address technological advance. No mystical force called 'the technological evolution' or 'the evolved human nature' is preventing us.

In the final analysis, the thoughts of Pinker and Gray on this matter are just two symptoms of technoparalysis. For both, the advance is a giant asteroid hurtling towards us, and the only real difference is in their attitude towards it: Pinker is not worried, while Gray is worried but feels powerless. Denial and resignation, basically. But it is not an external force beyond our control, it is a human activity. If that activity is to be changed, then we need to think about it. And for that to happen, we need a human race that is accustomed to thinking big – without frivolity, without impatience, and without a feeling of powerlessness. We have a tradition of that kind of thinking, with sub-traditions deeply rooted in all the main geographical regions of our world, and all containing enough wisdom to make this very obvious problem, which we need to do something about, as unavoidable as it should be. I suggest we use that resource. We do not need 'Enlightenment Now', as Pinker urges. The Enlightenment was only one movement in the history of philosophy and a mixed bag at that. What we could use is some philosophy now, so that gods and titans can both guide us.

6

Freedom

§1. An Anomaly

Freedom is highly valued within our world. With the possible exception of love, it is hard to think of anything more consistently highly valued. Money might be a contender were there not a certain stigma attached to it, but you cannot have too much freedom or love, it sometimes seems. People willingly die for freedom and are celebrated for doing so. In light of this fact, it is remarkable that the materialist philosophy has managed to catch on. For it is a philosophy which challenges the very possibility of freedom, with the most pro-freedom attitude it can muster being a valiant fight to save it against the odds – odds fixed by materialism.

Antipathy to freedom was not yet a theme in ancient materialism. Lucretius, probably following Epicurus, thought it could be accounted for by random swerves in the motions of the atoms.[1] But ever since materialism re-emerged in the modern period, inspired by the new mechanistic and mathematical science of Galileo, it has been characterized by either the outright rejection of freedom, or by efforts to preserve some modified notion of it. Baron d'Holbach took the former option, saying that, 'Man's life is a line that Nature commands him to describe upon the surface of the earth: without his ever being able to swerve from it even for an instant', and that, 'despite of the shackles by which he is bound, it is pretended he is a free agent, or that independent of the causes by which he is moved, he determines his own will'.[2] Hobbes took the latter, saying that freedom 'is the absence of all the impediments to action that are not contained in the nature and intrinsical quality of the agent'. But the illustration he then provides is that of a *river*, one which is free to 'descend, by the channel of the river, because there is no impediment that way'.[3] If our freedom is like that of a river, then Holbach could have said we were free too. These two options boil down to a linguistic choice over whether to describe the materialist vision as one which denies or accommodates freedom; whether to think of the natural forces which determine everything we do as 'shackles', as Holbach did, or insist that freedom can only be impeded by the real shackles humans use on each other, with Hobbes. This has remained the principal divide between materialists on this topic ever since; one of temperament, rather than discrepancy of vision.

The anomaly in our world between the supreme value we place on freedom and the background dominance of a philosophy that challenges its very existence, is something that could only have been made possible by anti-philosophy. For once people became inclined to thinking that science had all the answers, the implications were easy to ignore. They could continue to espouse the supreme value of freedom, while at the same time, if called upon, display their sophistication and lack of religious indoctrination by saying that reality is just the atoms following their natural course – with any apparent tension easily passed over as of only specialist interest. People can go along with materialism so long as it remains only background noise, for once philosophy is assumed to be isolated from life – if indeed it is even recognised at all – then they can carry on thinking what they did before. But what now just seems like words may, if left alone, reveal itself as the blueprint for the materialist reality we were building. I shall return to this thought in the final section of this chapter; first I want to try to solve the philosophical problem which materialism creates.

§2. Constraints and Restraints on Freedom

Freedom is being able to think or imagine whatever you like, and being able to move your body however you like. Freedom of mind and body are constrained by our abilities, which are in turn constrained by possibility. So, you cannot think about the life of Alcibiades if you have never heard of him and you cannot imagine a square that is also a circle. Similarly, only contortionists can wrap their feet behind their necks and nobody can fly. But within these constraints you are free.

Being able to move your body however you like is a very basic kind of freedom. Even a prisoner can do this within the confines of his cell. To have even this freedom restricted, by means of ropes, chains or, most dramatically of all, paralysing drugs, is the greatest violation of bodily freedom we can imagine. What the prisoner cannot do, of course, is walk out of his cell. But there are many places I cannot go too. I would be stopped long before I got anywhere near a military headquarters. Our freedom to do what we like is restricted by external constraints, some of which we may regard as reasonable, some not. Yet we can always try, so long as it is within our basic bodily capabilities.

Our bodily freedom is amplified when we interact with others and our external environment. Give me a Phillips screwdriver and I am free to drive in a cross-recess screw, just by moving my hand; give me a gun and I am free to kill a man, just by moving my hand. As our freedoms expand, the potential for them to impinge on those of others does too, and agreements must be reached as to boundaries where it is reasonable to curtail them. When agreement is reached, we are expected to show restraint, and if we do not, we can expect external constraint. Caring passionately about freedom is caring about where and how these boundaries are drawn; that they should not be drawn differentially by one class of people who has power over another is a point we have gradually been coming to some agreement over. The conversation is on-going and the technological advance continually feeds it. Our recently acquired freedom to publicly post messages, images and videos, and the freedom we feel we should have to engage in online life without being observed, analysed, manipulated, or harassed, is central to the current conversation. Freedom of technological innovation is not.

Just as paralysing drugs are the greatest violation of bodily freedom we can imagine, direct and non-rational intervention with our psychology, such as by means of neurosurgery, is the greatest violation of freedom of mind we can imagine. On reflection, however, it is not so obvious what this might amount to. In the former case you can no longer move your body as you like because something is stopping you from moving it at all, while in the latter, you can no longer think what you like because ... the natural parallel to draw here would be: something is stopping you from thinking at all. But then if that is what is going on, it seems less a matter of violating your freedom than of causing you to cease to exist.

To begin to see what is going on here, let us return for a moment to the bodily case, to consider an attempt made to non-rationally get you to kick a ball; non-rationally, in the sense that you are not being asked, bribed or threatened, but rather a brain-altering device is fired at you, for instance. In a clear sense, this could not succeed in getting you to kick the ball – you would not perform the action, because that would presuppose that *you* did it. What could be done, we may suppose, is that your leg could be moved involuntarily in such a way as to kick the ball – physically, the motion could be just the same. But you will immediately know you did not do it, just as if a doctor bashes your knee to trigger a reflex; it is a bodily event which will not 'figure from within the horizon of consciousness' as willed, as Valberg puts it.[4] And yet even if this same accompanying phenomenology could be caused by the device, you would still not have done it, even if you could never know this. The act would need to be willed by you, and although another might be able to induce in you the feeling of willing, nobody can will for you – only you can do that. In this sense, then, nobody can kick a ball for you.

Turning back to freedom of mind, then, let us consider an example Dennett has discussed. Tom, who has no siblings, 'wakes up to discover a non-rationally induced belief in his head': the belief that he has a brother in Cleveland. If it is just one isolated belief, Dennett thinks, then either Tom's rationality will be seriously impaired, because he will retain it in spite of its clash with his others ('I'm an only child with a brother', he might say), or the effect will be 'evanescent', because he will notice the oddity as soon as the thought crosses his mind and will not go on to endorse it. Dennett's conclusion is that 'one cannot directly and simply cause or implant a belief, for a belief is essentially something that has been *endorsed*'. This contradicts how he set up the example (Tom woke up with the belief), as well as his interim conclusions that Tom's 'induced belief can last only a moment', if he can remain rational, or that he may not be able to remain rational because it may be 'physically impossible to insert a single belief without destroying a large measure of Tom's rationality'. Presumably Dennett meant to say either that you can implant the belief, but it could not remain for long without seriously impairing rationality; *or* that you cannot, in which case Tom will never endorse the false information that flashes before his mind, or, if he does, then he will now be so devoid of rationality that he no longer believes anything at all – although in this case, you would then wonder what the endorsement amounted to. It is hard to know which was intended because the discussion points in all these directions at once.[5]

Looked at from the outside, you might indeed be unsure as to what to say, since there are various judgements an outsider could make as to whether Tom ever really believed it: perhaps he did for a second; perhaps he just said it and then wondered why; perhaps the lights are no longer on if he is now so irrational that he cannot spot blatant contradictions. But it is the situation on the inside we are trying to get at. It is convenient that Dennett had Tom wake up *after* the procedure because that was when the interesting subjective action took place. So let us suppose he was awake: what would happen? When we try imagining the subjective chain of events, it becomes clear that Tom cannot have been induced to believe something, because to induce a belief you would have to induce the 'I' having the belief – and it would not be Tom. Whereas in the case of kicking a ball we can imagine an alien force taking control of your leg, we cannot imagine an alien force taking control of your mind while you are still there to watch the show. To sense that a spirit has invaded my mind and is controlling my body, I need to be watching the show; if I only realise later, I was not there at the time. All that can have happened, then, is that Tom's stream of consciousness was interrupted to insert an alien belief, had by who knows what. Perhaps the idea of a momentary conscious self is nonsensical, since a moment of consciousness presupposes so many past moments, anticipated future ones, and conceptual mastery; perhaps it is possible, but Tom could not survive the disruption. But what does seem clear is that freedom cannot be removed from a subject of experience without removing that subject of experience. Similarly, you cannot non-rationally make Tom kick the ball, but only, perhaps, remove Tom from the equation by interrupting or terminating his stream of consciousness with another subject of experience who kicks the ball.

So, freedom of mind is essential to what we are; bodily freedom is not, because paralysing drugs need not end us. As such, we must reconsider how the former might be externally and non-rationally curtailed. You cannot directly induce an act of mind in this manner, but you could in principle affect the content the mind passively receives to indirectly influence its acts non-rationally. If, after having your brain manipulated, you suddenly seem to see a star, this is not fundamentally different from somebody placing a real star-shaped object in front of you – an experience has been caused for you to think about. If the aim is to make you think there is a star, this still counts as an attempt at rational persuasion because you have been given evidence. But if certain thoughts are encouraged by accompanying pleasure and others discouraged by pain, then you might be non-rationally conditioned to think certain things. This is non-rational, because although you are being given reason to think things and avoid thinking others, the reason, from your perspective, need have nothing to do with what you are thinking; pain may give you reason to curtail a certain line of thought, without providing any evidence against where it was leading. Brain manipulation is not essential to brain-washing, but it would be a quick and easy way to get the desired associations in place and ensure they were permanent. If 'moral enhancement' is coherently imagined, then presumably it would have to be of this kind.[6]

So, freedom of mind is the fundamental notion of freedom for two reasons. Firstly, because bodily freedom, and all the extended notions of freedom that arise from it as we interact with each other, is dependent upon freedom of mind – a bodily motion is not an action unless willed. And secondly, because our very existence depends on it – if my thought is not free to roam over my stream of consciousness, then I am not there and the stream is not mine. We need to restrain and constrain our freedom in order to live with each other, with the limit of external constraint for bodily freedom being the nullification of bodily control by an external agent, and the limit of external constraint for freedom of mind being the directing of thought through pain and pleasure association by an external agent.

According to determinism, however, reality itself places a far more radical constraint on our freedom than any external agent could ever manage. For determinism is the belief in maximal constraint, such that every last detail of your life is constrained to a T. It says that everything we think and do has been predetermined since the moment of creation . . . and then it puts the ultimate creation aside, once it has served its purpose of casting our own little acts of creation into doubt. On the basis of senselessness, then, all the sense we make of our lives is jeopardised, as determinism asks us to believe that we cannot think or do whatever we like, but must instead follow a predetermined course of action.

Can this count as 'acting'? Does a river act? Even Schopenhauer, an idealist who placed such stock in freedom that he tried to characterize transcendent reality itself as 'will', was so moved by deterministic thinking that he came to think of motivation as the subjective equivalent of causal determination, and so wrote that, 'Spinoza says that if a stone projected through the air had consciousness, it would imagine it was flying of its own will. I add merely that the stone would be right.'$_7$ Clearly, we are dealing with a powerful intellectual pressure to think that genuine human activity is an incoherent notion (but not the clockwork kind somehow already set in place). With everything active removed, we are left in search of a passive notion of action. We might find it by noting that we feel motivation and feel we are deciding, which a river does not. But then when we recall that materialism challenges subjective feeling too, a natural terminus begins to emerge: to deny that we exist. This could be reached by replacing our active sense of self with passive awareness of nature taking its course, then removing the awareness too. The result would be a description of a fiction, but fictions can be believed and acted upon. Descriptions of reality are a better basis for action, however, and that is what is provided by idealism, the philosophy of freedom.

§3. The Idealist Solution

The solution has three stages. The first requires us to see through the notion of causation in the subjective sphere of experience and thought; the idea that experiences might be caught up within causal mechanisms in the same way that physical things are. It is this materialism-driven thinking which makes it seem like free will is an incoherent notion, such that determinism no longer seems like the root of the problem anymore. It makes thoughts seem to inexplicably pop into your head, whether or not you suppose a neuroscientist could causally explain them from the outside, and the only escape from this artificially induced madness is anti-philosophy: 'freedom doesn't even make sense .... oh well ...' – and life goes on. So the second stage of the solution is to provide a more adequate philosophical reflection on our subjective experience of freedom. Then the third takes us to the root of the problem, which is the ancient dream of determinism.

Hold up one finger and take a look. Just concentrate on the tip segment where the fingernail is. Focus on your visual experience. You will find it has a size, shape and colour. We are not talking about the finger segment itself, but rather your visual experience – something which would *be* different if your eyesight was worse (the segment would just *look* different), something which will disappear if you shut your eyes. On the face of it, this has a size, shape and colour. But we know it cannot really, because then it would be part of the objective world. And we know that is not right, because when you change your viewing angle by moving your head to one side, you do not change your angle on the visual experience, allowing you to see another side of it, but rather change your angle on your finger and get a new visual experience.

So, your visual experience of your finger segment is not part of the objective world, although it seems to be, because it seems to have a size, shape and colour. That immediately tells you that it cannot be in causal contact with your finger. It cannot be that your finger is causing the visual experience, because for there to be causal regularities between your finger and experience, both would need properties like size, shape and colour, which the experience does not have.[8] So what is really going on? We are misrepresenting experience when we think of it as having a size, shape and colour, just as much as when we think of it as something with the potential to be in causal contact with things in the objective world. We misrepresent it as a potential occupant of the objective world in order to make the best sense we can of our phenomenal reality. Within our conception of phenomenal reality, objective things can be in causal contact with each other. But by raising the level to the metaphysical, we see that conceiving experience itself as phenomenal realities participating in this causal connectedness, as we must to maintain our conception of phenomenal reality, is incoherent. The incoherence is removed at the metaphysical level by reconceiving experience as transcendent. Then in light of this realization, our objective conception is reconceived as characterizing phenomenal reality only, while our experiences, which revealed this phenomenal reality to us, are reconceived as having always been a misrepresentation at the phenomenal level: a misrepresentation of the transcendent reality of experience. While our understanding of objective reality is self-contained at the phenomenal level, our understanding of subjective reality is not.

With this in mind, consider the case of free will. You think something. This thought is experiential, so it is not part of the objective world. So, in that case, it cannot have been caused by something in the objective world, any more than the experience of the finger segment can have been caused by your finger. So, there cannot be causal regularities connecting the thought with the immediately prior state of your brain, and from there back to a first objective event occurring some 13.8 billion years ago. This does not mean that the thought must have happened randomly, however. To say that something happened randomly implies a mere absence of causality. Only within a purely objective understanding could causation and randomness be a strict and exclusive dichotomy. Within subjective understanding there is also: *what I did* – and, as a consequence, within our objective understanding, *what he, she or it did*. Thinking that if a thought is not determined it must be random is like thinking that if a tree is not employed it must be unemployed – and then starting to wonder whether the state pays benefits to vegetation.

This brings us to the second stage. To conceive subjective experience in line with objective thought is something which materialism initially demands; 'initially', in the sense that for materialism to be a rational metaphysic, it would need to explain subjective experience in terms of the objective world. Materialism has not delivered on that demand, as we saw in Chapter 3, but its residue lingers on in the problem of free will through the idea that determinism and indeterminism are an exclusive dichotomy within subjective thought. On the materialist assumption that this dichotomy binds objective thought, it would have to bind subjective thought too, if experience were objective. So, it is materialism which has set this problem up, before fading away to no longer seem relevant.

If we accept the dichotomy, then free will suddenly seems like an incoherent notion – it seems that whether determinism *or* indeterminism is true, we cannot be free. The idea of determinism now seems tangential to the real issue because indeterminism would not even help: to think or do something randomly is not to do it freely. The real issue, it now seems, is with the very idea of free will. It is not that the truth of determinism makes it impossible for us to believe in that idea, but that the idea itself is incoherent. For once you think of experience as akin to the causal mechanism / random progression of the objective world, the illusion suddenly seems apparent to introspection alone – you need only reflect on your experience to see it. The choice seems to be between thinking that each of our thoughts, decisions and subjective bodily actions pops into existence inexplicably, like little Big Bangs, or that they occur inevitably as a consequence of previous events. Neither satisfies the notion of freedom as active choice between options. And so, in this manner, materialistic thinking has beguiled even idealists, like Schopenhauer; or, in the present day, an anti-materialist such as Galen Strawson, who cannot accept that we are unconscious robots, but feels compelled to accept that we are conscious ones.[9]

Sam Harris, one of the leading 'new atheists', crystallizes this thinking perfectly:

> Our sense of our own freedom results from our not paying close attention to what it is like to be us. The moment we pay attention, it is possible to see that free will is nowhere to be found, and our experience is perfectly compatible with this truth. Thoughts and intentions simply arise in the mind. What else could they do?[10]

So, for example, he has a coffee one morning, rather than tea. 'Did I consciously choose coffee over tea?' he asks. 'No. The choice was made for me by events in my brain that I, as the conscious witness of my thoughts and actions, could not inspect or influence.' This insight has nothing to do with materialism, he maintains, because the 'unconscious operation of a soul would grant you no more freedom than the unconscious physiology of your brain does'. To really be free, you would 'need to be aware of all the factors that determine your thoughts and actions, and you would need complete control over those factors'.[11] But this is paradoxical, he thinks. And perhaps it is, because he has just outlined a conception of freedom that requires omniscience and omnipotence, or something close; clouds are a factor that determine my thoughts about clouds, but I lack even the slightest control over them, let alone complete control. I would be very surprised if our ordinary conception of freely doing something turned out to be a theological conception of God's freedom.

Let us backtrack a little, to see how easy it is to avoid this predicament. We have already seen that we cannot, with any seriousness, think of subjective experience as part of the causal mechanism of objective reality. So how else can we think of our subjective freedom? Clearly not in terms of transcendent reality; characterizing reality as 'transcendent' ought to be enough to make anyone highly suspicious of further attempts to characterize it, and although the alarm bells rang for Schopenhauer, they were apparently not enough.[12] So if the causal connections which are only at home in our objective understanding are out, as is a direct characterization of transcendent reality, what mode of understanding is then left to us? We can and do understand freedom in terms of our social framework. Guided by desires and beliefs, but not caused by them, we choose our goals by creatively envisaging them, or by selecting from those that are available, or more often, a little of each. We make plans to achieve them, with consideration of the possible plans feeding back into the decision about the goal, and then we set out on a course of action, making adjustments along the way. As we progress along our chosen path, dealing with obstacles as we encounter them, and exploiting the causal relations we know of and can use, new options often come to light, which may lead us to change our minds, either about the plan of action or the goal itself.

The framework spreads over our lives in interlocking circles which widen and fade. A person's freely chosen project typically fits into a wider one, such that achieving its goal has a point within that wider project, the goal of which fits into a wider project still, and so on, with many of these projects feeding those of other people. This kind of framework understanding, of what we are doing in terms of what we are trying to do, is radically unlike that of a physical interaction which you might measure over a short period of time in carefully controlled conditions, so that extraneous factors are eliminated as much as is possible, thereby allowing us to say what caused what in a momentary transaction. Consider this example from Tallis about joining a pension scheme:

> I may visit an advisor one day, a few days later I get out the brochures he has given me and fall asleep over them, read them a few weeks later still, a

month or so on discuss it with my partner, visit the advisor again and so on until a decision has been made, then sign the various forms when it is convenient, and subsequently continue paying the premium while all the time reviewing it to see whether it is adequate for my changing needs or my perception of them. There are clearly large gaps in the sequence of events that end up with my drawing an appropriate pension when I have retired.[13]

Inspired by materialism, you might imagine that it must be possible to reconstruct this in terms of belief-states meeting desire-states to cause volition-states – because the states we have suddenly started talking about sound a little bit like atoms causally interacting. But now you risk losing your grip on framework explanations, like the one above, which we all understand perfectly well outside of philosophy, and can understand within philosophy too. That grip is liable to slide away from you in the course of your philosophical reflections, leaving materialism only able to restore sanity with anti-philosophy; as is vividly illustrated in the last two pages of Harris's book, where he becomes baffled over how he is choosing the next word to write, then ends by going off to have some lunch.[14]

The first thing to remember is what the framework is being applied to in the subjective sphere. Our way of being is that of body-subjects. Subjectively, that is how we are, and so far as we know, that is the only way anything is – objective stuff is not a way of being and neither is an event or state. You exist as a body known from the inside, which thinks, imagines and acts, and in so doing frequently loses sight of the body known from the inside. If you think of that body objectively, then that is not how you think of it *as you*, but how you think of the bodies of other people, and of your own body as experienced from another's perspective – photographs and mirrors help us to think this way. But as far as subjective thought goes, all there is to think about is the flow through time of a body-subject with a mind that wanders far over reality and possibility.

The framework provides us with an understanding of subjective activity which relates actuality to possibility, and derivatively, allows us to understand the bodily movements of others as objective manifestations of that same subjective activity. It allows us to think of a subjectively experienced bodily movement as *you doing something* in pursuit of a goal, and derivatively, of somebody else's physical movement as one explained by a certain goal which may or may not be achieved in the future. It is only when you try to think of this framework, project-directed understanding of our lives as an account of subjective reality to parallel our physical understanding of objective reality, that it starts to seem problematic; that is, when you try to think of it as a theory of the subjective nature of free will. But to do that is the big mistake which the initial impulse of materialism inspires. For the framework does not even aspire to tell us what exists within our phenomenal world – we have objective and subjective thought for that.

The framework makes no attempt to compete with the understanding of the nature of the subjective flow that we already have: a flow of experiences. This was borrowed from objective thought – it is its shadow – for we have no positive conception of experiences except that which is drawn from the conception of the objective world which those experiences provide us with. All we do is switch our interpretation of what we experience from, for example, 'the tree' to 'the experience of the tree', with each having different connotations; sometimes we need to get more creative to distinctively characterize experiences, but a switch in interpretation is at the root of the process (e.g. from something in your knee to a painful feeling). But unlike in the objective world, where our conception of what exists is content with objects in causal relations with each other, or else following a path we cannot predict from the causal relations we know of, in the case of our conception of a flow of subjective feeling we also have to contend with the fact that some of those feelings – an almost continual thread to the flow of the body-subject – are feelings of activity.

Whether active or not, a feeling is still a feeling. But to be true to active feelings, and thereby make coherent sense of them, we cannot seek to connect them up with others within the subjective flow exclusively by means of causation or randomness. We will do so where the phenomenology suggests it, for the subjective flow has both passive and active elements. For the passive, we borrow the notions of causality and randomness from objective thought. So we think of objective conditions as causing experiences, and of memories and images as sometimes just occurring to us from nowhere – although this subjective notion of randomness is different from the objective one, because it is the *reason* we are unable to find, not the cause. For the active, however, we have no alternative to the framework: to thinking of some elements of the subjective flow as what you, a free agent, are doing or trying to do. This spreads a logical structure over the subjective flow which *makes reference to that flow*, nothing more, and it does so in such a way as to do honour to the phenomenology of activity.

To see this, suppose we provide a framework explanation of a man's physical movement by saying that he was trying to do something or another. This refers us back to the subjective domain, because we are thinking of it as a willed action – so we might say that he did it because he desired something, for instance. But a desire is not a feeling. To make this point vivid, consider an intense, almost uncontrollable feeling of the need to urinate. That is not a desire, it is a feeling: it *gives you* a desire to urinate, we might say, as part of a framework rationalization loosely laid over the top of the feelings.

In the case of beliefs, hopes, decisions, and that vast body of logical points within our framework rationalization of the subjective flow which are not so closely tied to datable feelings, this is even more apparent. If you find yourself trying to focus on yourself believing, you will struggle: are you to focus on words almost heard in your head, or are you now engaged in a distinct, artificial activity, that was initiated in a failed attempt to recall the ordinary phenomenon? In fact you are not focusing on belief at all, but rather the feelings which that part of the framework loosely applies to – for 'belief' is what you say of someone feeling that way, or who has felt that way, or would. In applying the framework, we are not trying to name states or events whose nature might be further investigated, nor trying to provide a remotely accurate description of the phenomenology, which could never be done, for the reasons explained in Chapter 4. Rather, we are imposing a structure on the feelings which connects them up in a way we can understand. This structure applies to both objective and subjective thinking about what exists and is premised on our freedom.

Once it is seen that the framework is not competing with subjective thought by attempting to tell us the nature of feelings, but is rather providing a framework for connecting them in the way that it feels they ought to be connected – actively, in a great many cases – then there is no more reason to be suspicious about the coherence of the notion of free will. Causation does not provide us with a way of connecting the subjective flow which we can treat with metaphysical seriousness either, as we saw in the first part of the solution. So, we have no need to try to convert the nodes within our framework reconstructions into mental states which cause us to do things. And we should not, because the framework is always at one step removed from the flow.

So we do not have to falsify the phenomenology by collapsing the distinction between the parts of the subjective flow which feel active and those that feel passive by making them all passive, for if we should not take subjective causation metaphysically seriously, we have no reason to substitute it for the active framework alternative just because we cannot take that metaphysically seriously either. We have no more reason to deny that the phenomenology of the body-subject is active, by saying it only seems active, than we have to deny that an experience of a tree is passively received, by saying it only seems passive. Active and passive phenomenology distinguish different points within the web of objective thought, subjective thought and the framework – that inextricable trio with which we understand our phenomenal reality, and thereby make ordinary sense of transcendent reality.

So, consider Harris again. He thinks his thoughts 'simply arise in the mind'. This is because he is misinterpreting the framework in line with materialist metaphysics. He is thinking that free will requires one of his feelings – which is now what he thinks that a thought or decision is – must be the cause of his action, a cause he somehow controls. But when he looks inside to find that feeling, the isolated and objectified thing he finds seems like something he has no control over it. It is something that just happened to him, and he supposes that this is because his brain made it happen to him. What else could his thoughts do, other than 'simply arise in the mind', he asks. He could think them.

All free will requires is that the framework be cast over a variety of feelings constituting the subjective flow of a body-subject during a vague stretch of time that our framework understanding deems relevant on account of a particular project being deemed a significant subdivision of that flow. The project in this case was making a coffee, and the particular thought or intention which Harris very artificially looked for during this time, was his framework understanding applied to a moment in a string of active feelings in order to make it comprehensible to him as: me deciding on coffee and going to make it. The thought did not just pop into his mind, because understanding his experience that way was him thinking it. This is the understanding of active feelings – of me doing something, of me thinking something – which the framework provides. It makes what Harris was feeling at that time comprehensible in terms of his projects, those of a man currently in the habit of starting the day with either tea or coffee (one project) in order to perk himself up for a day working on his book (a wider one). It is no surprise that somebody in that situation would think 'I'll have a coffee'. If Harris did not think that thought, nobody did. And if nobody thought it but it did nevertheless occur, then he does not exist.

Notice how, to avoid the unbearable incoherence of actually thinking that he does not exist, he instinctively positions himself slightly off-camera, as the observer of the thoughts inexplicably popping into his mind. But the framework will also interpret the active feeling of that observer as: *him freely thinking something*, namely thinking that the thought just popped into his mind. If he had tried to disavow this understanding once again, by inventing yet another observer – an observer of his thought that the thought just popped into his mind – then he might have really 'lost it'. Without anti-philosophy to come to the rescue, materialism can be a menace.

Why did he choose coffee over tea? Because it was always going to be one or the other and he did not much care. But note what would have happened if he had suddenly felt a strong urge to drink some durian fruit juice, something with no precedent in his background – maybe he once tasted durian many years ago, maybe it is simply a fruit he knows to exist. The urge is so strong that he spends the whole day driving far and wide to specialist delis and shops, in order to buy some durian to juice. His brain would not have made him do that either, he would have done it. And it is because he would know this that he would not be able to dismiss the event by simply abandoning his confusing reflections to go off to have some lunch. He would spend days wondering why he did it; searching his mind for a reason, entertaining various hypotheses. In struggling to fit this episode into his framework understanding, he would have now found a real reason to be baffled by his own free will.

This would be a highly untoward event, of course, because our framework understanding generally makes our subjective lives comprehensible as, in part, me doing or thinking something, for a reason we would rarely care to specify but could if the need arose. This understanding is not meant to provide an account of the nature of subjective feelings which reveals how *they* get the job done, because it actually is the understanding each of us has of: *me* getting the job done. If you look to the materialist vision of events and objects in a causal mechanism for an alternative, then you rip your feelings out of the framework to leave them isolated, inexplicable, and no longer your responsibility – you become the bewildered spectator of your own mind, a spectator which, in all consistency, you ought to disavow too ... but cannot. So, you return to the framework, which your reflection has taught you to condemn. But it was not the framework that failed. The framework was your salvation, and with an adequate metaphysic in place, you need never have left it in the first place.

With free will now seeming more normal, I hope, we reach the third and final stage. Experience, the transcendent reality, is made sense of in terms of various streams of subjective experience, with us identifying with the active strands in accordance with our framework understanding, together with an objective world, with us identifying with a certain part as our body-subject considered from the outside. Even within this idealist picture, however, the problem of a clash between free will and determinism still arises, for the objective world is one in which we discover causal laws. If determinism is true, those laws determine what our future bodily movements will be. So, although we no longer need think that our thoughts and experiences are determined, nevertheless if all their outward manifestations are, little room is left for the subject to exercise freedom. At the very least, our framework understanding of bodily freedom, such that I could walk to the left or the right on some particular occasion, would have been shown to be illusory. So, although idealism explains the possibility of freedom, determinism remains a major obstacle to that possibility being actualized; but only if the objective world is a deterministic one, of course.

So, what could persuade us of determinism? Either an *a priori*, reason-based argument, or an empirical, experience-based one. An *a priori* argument for materialism would not be enough, even if there were any good ones, because materialism is compatible with indeterminism: the material reality might not be completely governed by deterministic laws. Helen Steward, who is a materialist, is quite sure that it is not: by the end of her book on free will, she has decided that we live in an 'unquestionably indeterministic world'. She describes the determinist's view that 'wars, stock market crashes, global warming, revolutions, industrialization [and] the myriad small decisions each of us makes on a daily basis' are simply the 'high-level manifestation of the inevitable workings-out of the consequences of the initial conditions at the start of the universe' as being 'perhaps one of the most astounding things that has ever managed to obtain the status of philosophical orthodoxy'. I agree, and will try to explain how it happened in the next section. The root of her conviction is that we are free, but as a materialist, she sidelines the issue of conscious experience to understand this objectively as the claim that animals are 'self-moving' objects. Given the contentiousness, by her own admission, of her subsequent efforts to outline a 'top-down' conception of causation which would allow the causal powers of animals to be greater than the sum of their parts, and her modest conclusion that the idea is at least coherent, it is not clear she is entitled to such strong conviction. But nevertheless, indeterminism is certainly up for grabs for materialists.[15]

We need not wait for a clinching argument to be discovered as determinist and indeterminist materialists clash in accordance with Smart's 'standard picture' (see Chapter 4), however, because causation is an explanatory relation, not a physical one, and as such, it refers us back to the framework and subjective thought. Causal relations are posited between things that human beings want explained, by reference to what we consider relevant. Strike a match and we naturally consider the striking as the cause of the flame, but a Venusian might just as naturally think it was the oxygen in Earth's atmosphere; maybe they strike matches all the time back home and they never set alight.[16] But determinists do not want the interest-relative, explanatory notion of causation we actually have, they want objective causal powers that were built into reality at the moment of creation, such that everything was necessarily connected thereafter; something built into the 'real essences' which Locke said we could never know, producing the 'necessary connections' Hume found he had no experience of. It is an oddity that determinism has thrived within the empiricist tradition of philosophy.

We can try to get an objective notion of causation, however, by artificially defining one; J.S. Mill did this in the nineteenth century with his notion of 'total cause'.[17] A total cause is the sufficient condition for an event to occur. This might have to include everything in the prior state of the universe, for all we know; in current physics we could restrict the state to a sphere large enough that nothing outside that sphere could send a signal to something within it without travelling faster than the speed of light, but as we learn more about the 96 percent of the physical universe we currently estimate we know nothing about, this could change. If it is not already preposterous enough to imagine that we might ever come to know causes of this kind – the kind which determinism needs – then add the fact that we certainly could not unless they had nothing to do with us, which might not be possible. For if our 'knowledge' was part of the total cause, then the cause would be altered and the knowledge negated.

But the notion of total cause was fabricated from our explanatory notion of cause, and still looks suspiciously explanatory – total causes *would* explain these cosmic transitions, if only we could grasp them. And if it is explanatory, then the basis of the determinist's attempt to overthrow our perfectly adequate framework explanations of what people and other animals freely do is simply a vision of an alternative kind of explanation. Explanatory notions refer back to human practices and all of our practices are weaved around the framework, which idealism can relate to metaphysical understanding.

Even if the notion of a total cause did successfully abstract from human interests to lead us to a purely objective feature residing in enormous, unfathomable chunks of physical reality, the prospects of our ever knowing them are negligible, even on the dubious assumption that we could peacefully wield the power accumulated on the path to that kind of knowledge for long enough to attain it. Thinking the universe transitions from one state to the next does not require total causes, because the transitions might be indeterministic. Indeterministic transition might mean nothing more than: in a manner which is compatible with our framework understanding, but which looks random when we take a purely objective point of view. So, there is no prospect of an *a priori* proof of determinism. The closest we could get would be that determinism is possible, a possibility already suggested by the artificial language of 'states' and 'transitions' that the determinist favours.[18] In any case, what contemporary indeterminist and determinist materialists are arguing over is causation, not total causation – about whether causal relations between atoms provide our best objective explanations of the motions of large-scale conglomerations of those atoms, or whether it sometimes works better the other way around – so their debate is orthogonal.

What about an empirical argument, then? Nobody has ever experienced a total cause. But if we put that absolutely crucial scruple aside, and also ignore our ubiquitous experience of freedom, then we might suppose that physics provides our best guide to the causal relations in objective reality – rather than the best causal explanations at the microscopic level. Since the 1920s, our best physics, quantum mechanics, has described the world as experienced by observers, not the ultimate, observer-independent objective reality of determinist materialists. It tells us that we cannot simultaneously know both position and momentum: we can know where an electron is but not what it is doing, or what it's doing but not where it is. It can tell us the probability of a particular nucleus decaying in a certain time period, but not whether it actually will decay. Quantum mechanics can be given either an indeterministic or deterministic interpretation that is consistent with the observations, but the deterministic one only works by assuming there are various details we are ignorant of; an assumption based not on experience, but either on the metaphysical assumption of determinism, or simply as methodological encouragement to new investigations.[19]

Still, it may be true nonetheless. All kinds of highly unlikely things are possible, so let us explore this possibility, since I already conceded that if our bodies are determined, the idealist conception of freedom is all but nullified.

The idea that knowledge of the deterministic causal laws could tell us exactly what our future bodily movements will be first needs some clarification, however. Such knowledge might tell you what somebody else's future bodily movements will be, so long as you are only a voyeur who will not interact with them; an astronomically distant voyeur, if we are talking about total causes, as we should be. But whether it could tell you what your own bodily movements will be depends. It depends on whether you have the character of a Greek villain, like Pelias, who tried to avoid his fate by sending Jason on the expedition for the golden fleece, or rather the character of a Greek hero, like Idmon, who accepted his fate by joining the expedition, knowing he would never return. If you are like Pelias, then on reading the prediction, you will, by the exercise of your own free will, falsify it with ease: whatever it says you will do, you simply do not. But if you are one of those rare souls like Idmon, you will follow the prediction. This might be extremely hard to do, however – practically impossible, if the prediction is exact. Suppose you saw a video showing your exact movements on the way to make some coffee. You would never get that exactly right on a first attempt, since your free will would get in the way. Only an Idmon-like robot could do it.

Nevertheless, although I could not be told the deterministic prediction in advance, I might be shown a computer-generated video of what it was predicted I would do an hour ago, alongside a video of what I actually did. If the match was exact, then freedom is an illusion. This is the kind of empirical argument that would work, so when I see that evidence, I will accept that determinism is true.[20] Until then, I will assume that when it seems to me that I could walk either left or right, then our best understanding of the objective world will be compatible with either of these outcomes. Of course, if I had overwhelming motivation to choose one of these directions rather than the other, then it was always obvious what I would do; but that does not mean I could not have done otherwise. I could have thrown myself over the cliff even though I had no reason to do so – possibility is just as cheap in the subjective realm, as you reflect on what you could do but never would, as it is in the objective, when you reflect that Elvis Presley may still be alive, for instance. But we do often have genuinely competing motivations, the framework tells us that things could have gone either way, and determinism only provides reason to doubt this if you silence your philosophical and scientific scruples for step after step after step, in order to imagine an experience there is every reason to think that nobody will ever have.

§4. Astrology and the Metaphysics of Desire

J.T. Ismael, another contemporary materialist trying to defend freedom, comes to the same conclusion about the deterministic dream of predicting exactly what we will do before we do it. She says 'you can laugh in the face of anybody who tells you she can predict your choices with certainty, because as soon as she tells you her prediction, you can undermine it'. She goes on to say that insisting the prediction could in principle be made, even though you could never be told it, is like 'a heavyweight champion who remains champion only because he never meets Ali in the ring.'[21] But suppose a fight with Ali is arranged for a future date and the champion studies what is determined to happen very carefully. He notices a point at which Ali makes a move which would, had the champion anticipated it, have left Ali exposed to a knockout punch. Now he does anticipate it. True, he will struggle to play his part well enough to reach that point in the fight without changing Ali's behaviour too much. But he might just pull it off.

And here we see the real allure of determinism: power. To know the future is to have power over those who do not. This is an idea which has always obsessed people. In Palaeolithic times, we recorded lunar cycles on animal bones and the walls of caves, in part for practical, time-keeping purposes, but there was probably also mystical interest, since by the beginning of recorded history, the astrological quest for meaning in the sky was already well-established. For most of our history, powerful men have sought astrological guidance when making important decisions. Astrological traditions have emerged all around the world, sometimes completely independently, accompanied by complex astronomical systems of knowledge for tracking the movements of the stars, since these movements and alignments, according to the fundamental premise of astrology, are what determine our individual and collective fortunes. Astronomy was once valued, in large part, as a resource for astrology. Ptolemy, the most influential astronomer in the West until the scientific revolution, wrote not only his *Almagest*, the basis for over a thousand years of astronomical research, but also one of the most influential works in the history of astrology, the *Tetrabiblos*.[22]

Astrology often came into conflict with Christianity, a faith which accorded all power to God. God could know the future, but it was impious to suppose that man could. This is the line Saint Augustine took, dismissing astrology as 'a delusion'.[23] But he did not think God's absolute foreknowledge ruled out our being free, because he thought our future free acts could simply number among those things God knows; on the face of it, this position was an early forerunner of Hobbes' compatibilism of 'free' rivers. Not all believers saw it this way, however. Procopius, about a century later, takes the rather more straightforward, unfree-rivers line that anticipates Holbach's incompatibilism, by saying that, 'it is not our own devices that control our lives but the power of God'.[24]

Astronomy was the trigger for the scientific revolution when Galileo's observational science of the stars came into conflict with Ptolemy's astronomy, which by this time was Christian orthodoxy, since it was earth-centred, and hence placed us at the centre of God's creation. Galileo was a practicing astrologer, who was not only put on trial by the church for opposing Ptolemy's astronomy, which is the famous story, but also, at an earlier time, for his astrological predictions.[25] But astrology, by means of its handmaiden, astronomy, was not destined to win the day over religion by cultivating the magic of modern science. Quite the contrary, it found itself an even more powerful enemy, which quickly revealed it to be the superstition we know it as today; it was already in steep decline by the end of the seventeenth century.[26] But it left its mark as something that is still considered a rational belief: determinism.

Hobbes, the father of modern determinism, was greatly influenced by Galileo's mechanistic approach to nature, which he encountered in person when he visited Galileo on one of his European tours.[27] Hobbes was inspired to take the same approach to people and society within his political philosophy. And although he was no fan of astrology himself,[28] when he combined his materialist philosophy, according to which people are just another material part of nature, with the idea of using observational science to predict how matter will behave, the old astrological dream resurfaced in a more philosophical guise. We were no longer to look to the stars to divine the fortunes of men, but to the atoms from which the stars, and everything else, is made. By making this transition, the new astrology insulated itself from empirical falsification in a way its predecessor could never manage. Old astrology has always had to make do with the defensive measure of stating its predictions in maximally obscure language. But now, instead of holding that our futures are determined by the alignment of stars at our time of birth, which, being knowable, made astrological predictions vulnerable if they ever became too precise, the new astrology appealed to the alignment of atoms billions of years before anyone was born. This alignment was suitably remote from human knowledge and the prediction was now even vaguer: simply that something or another is determined to happen. These extra layers of protection allowed the old superstition to survive into the present day as a respectable belief.

Now that the predictive powers of science have become so manifestly powerful, the original astrological dream of issuing actual, accurate predictions of our futures is resurfacing, secure in the conviction of deterministic materialist philosophy that such predictions must, in principle, be possible. Empirical psychology, now conceived by its leading practitioners as a natural science, is increasingly focused upon our brains, and if the causal connections between brain states and subsequent bodily movements can be fully mapped out, that raises the prospect of our reasons, choices and personalities being bypassed entirely to allow humans to be understood as predictable objects. Educated people now favour news of the latest discovery about what our brains make us do over horoscopes, but in the absence of portable brain-reading devices, a deterministic universe, and Idmon-like character, they are of similar use. Nevertheless, to actually have such predictive capabilities would yield incredible power. All the old dreams of the warriors and merchants who visited astrologers would have come to fruition for their counterparts of today. And if the prerequisite knowledge proves too long in the coming, given that we are not so good at predicting earthquakes and volcanic eruptions at present, there is always the idea of persuading us to change ourselves into more predictable objects. The more post-human we become, the more humanly-designed we become, the more predictable for the designers we become. Those who find freedom a burden – and making decisions can be stressful – will find others more than willing to remove that burden.

Materialism has always encouraged us to forget ourselves to focus exclusively on what we can manipulate, since manipulation gets us what we want. The manipulatable becomes reality itself in this metaphysics of desire. It is a testament to the goodness of human nature that its influence has been exerted alongside improvements in the lives of the many. But being an incoherent and largely unnoticed philosophy, the few driving the process cannot be expected to apply it consistently to themselves. Officially disavowing your own ego is no obstacle to revelling in the thought of what the future inevitably has in store for you, so long as it is a future you want; and a belief in inevitability need not lessen your efforts to make it happen, but may rather spur them on – just as astrology has always spurred people on, sometimes to victory, sometimes to disaster. Without knowing what future is determined, determinism for the individual is just a tool to persuade others of the pointlessness of opposing your efforts, and, if you are able to believe it yourself, to encourage those efforts and provide consolation in the case of failure. But if those efforts are directed at the collective, determinism can be an instrument of social change.

The bare supposed fact of it, absent the knowledge of the future it promises, might be enough to persuade us that we should reform our penal justice systems, for example, on the grounds that we ought not to be punishing people for committing crimes they were predestined to commit billions of years before they were born. Such thoughts are just as irrational at the collective level as they are at the individual, not just because they are based on determinism, but also because if determinism is true then it applies across the board – so without actual knowledge of the future, nobody can know if these reforms are destined to help or hinder us. You may think they will help, given the truth of determinism, but perhaps it is predetermined that when human societies try to organize themselves around the deterministic nature of reality, at a point at which their knowledge of that reality is highly imperfect, then those societies will break down. Nobody knows, so we just have to decide – but now a false view has influenced our free decision.

Anti-philosophy is likely to kick in long before determinism will ever influence the public sphere quite so directly; Clarence Darrow's famous determinism-based legal defence remains only an amusing anecdote. But indirectly, through the more concrete dream of actually predicting the future, determinism has a much greater chance of influence, since this allows it to rely on science and technology, and thereby leave its philosophical nature in the background. So long as we live recognisably the same kind of lives we have lived for thousands of years – going in and out of buildings, working, playing – it will be a great struggle to improve on the predictions of our behaviour that can be made using the framework of reasons, goals and choices. The desire to make more reliable predictions by treating us as conglomerations of atoms will always face the problem that the conglomerations are vast and complex. But the more we are monitored, the better the predictions will become. And if we increasingly live our lives in an artificially designed environment which can be fully monitored, we become more predictable still. This is a place where the goals available to us can be carefully crafted, our options for trying to achieve them already considered, and our choices predicted from a careful analysis of our past choices. If we were able to merge with technology, becoming artificially designed ourselves, we might be completely predicted, like a clock. We may never get very far down that path, but it is one along which there will be plenty of power to be seized. If we do not like this direction we can choose another one, and it would help if we stopped thinking we are determined. And for that to happen, as many people as possible must consciously entertain the thought that we are not.

7

Soul

§1. Making Souls

Is the word 'soul' worth saving? We have had it since the Old English of *Beowulf* and it is in no danger among today's growing religious population. It took on new life in twentieth century music as soulful styles developed in both jazz and rock, and there is much concern today about soulless corporate culture, architecture and art. Although atheists may be happy to use the word metaphorically, however, and in a sense in which soulfulness is an unmitigated good, most would disparage any suggestion that there really are souls. The religious connotation is simply too strong: souls survive the death of your body to continue into an afterlife. Atheists cannot accept that, and combined with the popular objectified imagine of a soul as an invisible human, or a cloud floating near your brain – i.e. a dubious part of the physical world conceived by forgetting the subjective thought behind the concept – the soul becomes an obvious target for scorn. 'Mind', a much less controversial word, comes from the Greek *menos*, which has the meanings of life-force, desire and spirit. When Homer's gods want to rouse their heroes in battle, they give them *menos*. 'Soul', on the other hand, is an English word without a Greek or Latin root, but which has traditionally been used to translate the Greek *psychē*, the root meaning of which is breath. When Homer's heroes die it is their *psychē* that departs for Hades, rather as, in *Beowulf*, souls depart from bodies to head for hell. Nevertheless, *psychē* gives us 'psychology', a study of the mind, not soul.[1]

The reason 'mind' and 'soul' have now acquired a certain independence seems to be primarily a result of Aristotle's theory of *psychē* and Descartes' reaction against it. Aristotle rejected Plato's conception of *psychē* as an autonomous being with only a contingent connection to its body, and replaced it with the notion of a form which animates our bodies and accounts for life in general, such that animals and plants have souls too, albeit more primitive ones than us. For Descartes, however, a living body was a machine, and its life was to be explained by mechanical principles. He wanted to return to a more Platonic conception of *psychē*, but given the association 'soul' had with the dominant Aristotelian tradition of his time, he chose another word, 'mind'. Although not consistent in his usage, Descartes made his intentions clear when pressed: 'the substance in which thought immediately resides is called "mind". I use the term "mind" rather than "soul" since the word "soul" is ambiguous and is often applied to something corporeal'.[2] He explains this ambiguity as the result of 'primitive humans' failing to distinguish the principle by which we think, from that by which we are 'nourished and grow and accomplish without any thought all the other operations which we have in common with the brutes'; these operations were to be explained mechanistically, for Descartes. So he said that, 'I consider the mind not as a part of the soul' – as Aristotle had thought of the rational, thinking soul as a higher part of a complete soul, the other parts of which accounted for the life of the organism – 'but as the thinking soul in its entirety'.[3]

So you might say that Descartes went for 'mind' rather than 'soul' because the connotations of 'soul' were too physical for him; it was the mind that was only contingently related to the body for Descartes, and which left the body at death. Then again, given that he says that the mind is 'the thinking soul in its entirety', perhaps his theory of mind was just a new theory of soul. In light of this history, then, I see no reason to exclusively annex 'soul' to religious views, such that atheists may only respectably speak of 'mind'. The manner in which the two became separated is that Descartes used 'mind' to replace the Aristotelian conception of *psychē* with a broadly Platonic conception of *psychē*; and 'soul', not 'mind', has been the standard English translation of *psychē* ever since the Greek of the New Testament was first translated. Descartes had a formative influence on modern philosophy, so 'mind' became the dominant word in philosophical discourse, secular conceptions of mind developed, and these days 'mind', unlike 'soul', does not immediately suggest survival after death; the religious may believe that our minds outlive our bodies, but this is an optional addition to the core idea. But in this regard, life after death could just as well be regarded as an optional addition to the core idea of 'soul', to make room for secular conceptions of our mortal souls.

There are a number of reasons why I find this idea attractive, but the main one is that a core component of the idea of soul has always been that it is what each of us, deep down, really is. That, of course, is why people want their souls to survive the death of their bodies. To say that *what I am is a mind*, on the other hand, is much less appealing, because 'mind' has connotations of pure intellect, which is only one aspect of our inner lives, and not one which many of us would want to fully identify with. Descartes used a revolutionary conception of 'thought' to encompass much more than just 'thinking' as normally understood – i.e. believing, following a train of thought, reasoning – such that bodily sensations and emotions were also 'thoughts' for him. But although the philosophical tradition followed suit in a manner, such that thoughts (in the normal sense), emotions and sensations are standardly all lumped together as 'mental states' by philosophers these days, the intellectual connotations of 'mind' remain strong. To say of an elderly couple, married for over fifty years, that they have come to know each other's souls, seems quite different from saying that they know each other's minds; a good confidence trickster might, in short order, come to know their minds just as well, perhaps better. Given this wider context, then, to use 'mind' rather than 'soul' would be to load the dice against what I think is the correct metaphysical view.

There are other benefits too. Since the word 'soul' has been neglected by the philosophical tradition, it has been spared the influence of materialism. Standard materialist talk of brain states 'realising' mental states takes us far from the idea of the mind as a substantial unity, which was Descartes' idea. But unity remains integral to the idea of soul, and unity is what is needed to address the question of what we essentially are. Moreover, a commitment to souls, as opposed to minds, is blatantly anti-materialist. And although this is largely explained by the religious connotations, the resonance could be useful for distancing us from the culture of materialism and scientism – a world in which atheists were relaxed about souls would surely have progressed far beyond that culture. Reclaiming the idea of soul might also help to break down barriers between the religious and non-religious – a task that may prove of growing importance to our world – since a disagreement about nature is less radical than one about existence. And it would be a natural development from the new, more secular, and essentially positive discourse of soul that developed in twentieth century mainstream culture, which is all about expressing and revealing what you really are. Philosophy would be relating it to something real. So I shall say 'soul' rather than 'mind' from now on.

Now these days, the essentially materialist idea of the soul as something you can *make* is starting to have real life consequences. People have been getting together to make souls for quite some time, of course, but the idea of souls as a kind of artefact you can produce to varying designs and with various different materials is something new, at least to the extent that we have learned to think of them this way ourselves; perhaps we always imagined that our gods did. Philosophically, the crucial turning point in bringing this about was Locke's discussion of personal identity, added to the second edition of his *Essay Concerning Human Understanding* in 1694. The famous thought experiment from that discussion is as follows:

> For should the Soul of a Prince, carrying with it the consciousness of the Prince's past life, enter and inform the Body of a Cobler as soon as deserted by his own Soul, every one sees, he would be the same Person with the Prince, accountable only for the Prince's Actions: But who would say it was the same Man? The Body too goes to the making the Man, and would, I

guess, to every Body determine the Man in this case, wherein the Soul, with all its Princely Thoughts about it, would not make another Man: But he would be the same Cobler to every one besides himself.[4]

Note two things. Firstly, it is the prince's *soul* that carries with it the consciousness of his past life – his memories and the personality he picked up in the life he remembers. The body of the cobbler becomes the body of the prince because soul-transference takes place, so on the face of it, you might naturally conclude that the prince, deep down, is really a soul: his soul now animates a new body. But what Locke actually says is that 'he' (the body of the cobbler) 'would be the same Person with the Prince'; by which he means that the body of the cobbler after soul-transference would be the same person that the body of the prince was before soul-transference. The section ends with Locke noting that 'in the ordinary way of speaking', the same 'person' is just the same 'man', but that we are all free to use words as we see fit etc. etc.; essentially, he admits to having just coined a new way of using 'person', rather as Descartes did with 'mind'.

Secondly, note that immediately after saying that the cobbler man/body 'would be the same Person with the Prince', he talks about *accountability* - the cobbler man/body would now be accountable for crimes committed by the prince man/body when it was occupied by the prince's soul.[5] Locke says that 'person' is a 'Forensick' (legal) term, as is often noted and usually thereafter ignored, and in a letter to William Molyneux, at whose request the personal identity chapter was added, he explained that, 'I am shewing that Punishment is annexed to personality, and personality to Consciousness.' In fact, one of Locke's leading concerns seems to have been to show that on Judgement Day, a just God would not punish us for crimes we could not remember, and hence for which we could hardly ask for forgiveness; according to his new coinage, because the person awaiting judgement would not be the person who committed the crime.[6]

This is a far cry from a metaphysical account of what we essentially are. Locke's philosophy emphasises, above all else, the limitations of human understanding. These limitations prevent us from knowing the substance of reality, he thinks, so since souls are substances, we cannot know their nature. Nevertheless, Locke finds it 'the more probable Opinion' that the consciousness determining the identity of 'persons', in his sense, is 'annexed to, and the Affectation of one individual Substance', i.e. a soul. And while conceding that his theory might look 'strange', since it holds that the identity of persons has nothing to do with the identity of substances, he thinks it 'pardonable in this ignorance we are in of the Nature of that thinking thing, that is in us, and which we look on as our *selves*'.[7] In acknowledgement of our limitations, then, Locke aimed to provide a practical account of our identities for use in judicial matters.

Despite the modesty of this proposal, however, some of Locke's contemporaries sensed the stirrings of an assault on our very identities; that 'personality is not a permanent, but a transient thing: that it lives and dies, begins and ends continually', as Bishop Butler put it.[8] And as it transpires, they were exactly right – for the discussion Locke inspired was eventually to bring us to the claim, currently considered respectable, that we do not exist.

The definitive twentieth century 'Neo-Lockean' account of personal *survival* – no longer identity – was due to Derek Parfit.[9] Survival replaced identity to account for variants on Locke's thought experiment where the soul of the prince enters two different cobbler-bodies at the same time; this is imagined as the result of cutting the prince's brain in two. The resultant people in the cobbler-bodies cannot both be identical to the person who was in the prince's body, so the prince-person is instead said to *survive* as two people. If Locke's account were still on the table, this might seem reasonable enough: perhaps both of the cobblers ought to be punished for the crimes of the prince. But it had by this point morphed into just one consequence of a metaphysical account of our non-existence.

To see this, consider Parfit's own famous thought experiment of 'teletransportation', or teleportation. A teleportation machine makes an exact physical replica of a human being in another place. This idea is a mainstay of science-fiction, as well as a research programme in contemporary science, insofar as there is already great interest in teleporting particles, and some scientists speculate on whether, or when, we will be able to go all the way.[10] As teleporting is usually imagined, the moment the replica is created the original is destroyed – or, as would normally be said when a human being is destroyed, killed. So, one human disappears, another indistinguishable one appears in a distant place, and the illusion of travel is created for onlookers. But that is not what Parfit thinks. He thinks personal identity is the illusion, and that since the relation between the original and replica is one of survival – understood in the Lockean manner of a continuity of consciousness – a person has travelled. As Parfit makes clear, the destruction of the original is inessential to the main idea, for once we have abandoned the idea of identity in favour of survival, whether the original is killed makes no difference to whether the person who teleports survives.[11] If the original is killed, the teleporter survives only as the replica, and if not, as two people: one who has travelled and one who has not.

Travel is also inessential to the main idea, so let us imagine a case where a teleporter makes a copy to sit beside me in my office. It leaves me intact, simply scanning my physical constitution to create the replica, who looks and acts exactly like me. Then the evil genius who made all this possible hits me with an excruciating pain caused by his brain stimulation gun, and tells me that if I press the button on my desk, the pain will be transferred to my replica.

If Parfit's position were correct, then I ought to consider it a matter of complete indifference whether I press the button or not, since the replica has an equal claim to being the person I urgently do not want to be feeling excruciating pain.[12] *Everybody* would press the button, however, which shows that once thought through, Parfit's theory is impossible to believe; to recognise that in circumstances most testing to your belief in the theory, you would act on the assumption that it is false, is to recognise that you cannot actually believe it. But I have loaded the example by presupposing that the original human is the person with the pain and the decision to make, of course; if Parfit were right, I might, with equal justification, have described the situation of the replica, looking over at the original and hoping he will choose to keep the pain. Once the replica comes into existence, two people have survived my life up until that point, and it is only a false belief in personal identity that makes me want to identify with one or the other, according to Parfit. But that only makes sense if I never really existed in the first place.

Suppose I did exist prior to this event. In that case, whether I am a soul, a body, a union of the two, or even just a stream of conscious experiences (the idea which making the mistake of interpreting Locke metaphysically suggests), there is still no reason to think the sudden appearance of something exactly similar would have any effect on me. The appearance of such a stream of experiences on Mars, for example, would obviously not mean I would suddenly find myself looking out over a Martian landscape, any more than, to take an objective parallel, the appearance of a replica of my desk on Mars would affect my desk; cover it with Martian dust, for instance. All that has happened is that a new individual of the same type has been created. But if I did not exist prior to the event, then the theory suddenly makes sense. I never was a stream of conscious experiences, in the sense that the particular connected experiences I had throughout my life was what constituted my existence as an individual; an individual object or an event or process. Those experiences existed, but their combined existence is not the person. The person is that *type* of event. This explains why all that is needed for the person to survive is that experiences of the right type continue to exist.

In a sense this is obvious.[13] If a person was an individual being, then it would not be able to survive the destruction of any particular individual being, as a person can on this theory. But it could as a type. Whatever I am, however, I am not a type. I have this much in common with my desk: we both exist as individuals if we exist at all. Perhaps you could make lots of people just like me – people of the same type – just as lots of desks like mine have been made. But my desk is not a type and neither am I. So on this theory, I cannot be a person, in the sense that this is what, deep down, I really am. Rather, I simply belong to a person-type, just as I belong to other types, like being brown-eyed. Neither can I be a person in the sense that this is the essential type I belong to, the one that makes me the individual I am, because various different individuals could belong to this type too. But the point of discussing survival in fantastical scenarios, now Locke's judicial interests have been side-lined, is to determine what core of our being must be preserved for us to avoid oblivion. So since I cannot be a person – because a person is a type – and being a person is presented as the best account of what, deep down, I really am, the upshot of this theory is that I do not exist; and neither do you, and neither does anyone else.

This is the conclusion materialism encourages, as I have mentioned before. It comes out only slightly differently when eliminative materialists consider a temporal freeze-frame of the soul and find nothing there, to when metaphysicians of all stripes, but against the backdrop of materialist influence, consider the soul enduring through time and find nothing there. It results from trying to side-line subjective thought, which is an essential component to how we make sense of reality, and the one which brings each one of us into the centre of the picture. And yet when the conclusion that we do not exist is actually reached – and in this area of philosophy, it comes as close to being explicit as it ever does – then rather than recognising this result for what it is, namely quite possibly the greatest example of a *reductio ad absurdum* which human intellectual culture has ever produced, it has somehow become possible to regard it as a surprising and interesting truth: 'Well I never, the self is an illusion . . .' What should really follow next is: 'It turns out I don't actually exist,' but somehow it never does.

The idea of souls as types suits crafting and manufacture. Just as types of cars are developed, with popular models evolving to overcome the problems found in their predecessors, such that updated variants emerge within what is recognisably the same general type, the same could be done to souls if they were objectively specifiable types. Once subjectivity is out of the picture, there is little difference between the continuity in style between a Mini Cooper of the 1960s and 2000s, and the continuity of behaviour between Minnie before and after she is modded with superintelligence. You can upgrade individuals too, of course, but you can only go so far before so little remains of the original that you might as well have started from scratch, as if you tried to turn an individual 1960s Mini into a 2000s one. Types are far more malleable.

Once you have designed a type, the individuals can be manufactured in any quantity required. And best of all, types are immortal. If no new Mini Coopers are built for a thousand years it makes no difference to the type, which could be built just the same after the hiatus. Human hopes for immortality in our technological world follow just this model. People have been paying for cryonic suspension since the 1960s and the numbers are rising (if not yet to significant levels since the costs remain very high), while others place their hopes for immortality in having their personalities uploaded into computers. Both hopes presuppose that persons are types, for to bring an individual back from the dead is just to make it a living type again – with the relevance of the future resuscitators using that particular corpse dubious in the extreme – while a personality that can be run on various different computers is evidently, like any programme, a type. If we never existed in the first place, we cannot die. Combine fear of death, faith in technology, materialism, and a good dose of anti-philosophy (so that none of this gets reflected on too much), and the mix is so potent that some people can be led to believe anything.

§2. Ego Death and Substance

Death is most frightening when considered subjectively. To see an image of your own corpse would be disconcerting mainly for what it signified subjectively. Similarly medical definitions of death, which change as we learn more about the human body, are primarily of interest because they provide our best objective estimate about the death we are trying to track in the subjective domain, namely the complete and permanent cessation of conscious experience. Given that fear of death is part of the human condition, it is understandable that efforts have been made to console us. The most popular in our history is to deny it ever happens through doctrines of the immortality of the soul. However since this has often gone along with the threat of punishment in the afterlife, which builds on and accentuates the fear, the ancient materialist reaction, best represented by Lucretius, was to accept death but try to intellectually undermine our fear of it.[14] To this effect, he argued that it is no worse than the nothingness that preceded our births; the flaw to which is that the prior nothingness never threatened us with complete loss. The denial of individual existence takes something from both these strategies to generate an incoherent one. It promises immortality, while claiming that death is nothing to fear because we never existed in the first place.

For such a tactic to have any success, our certainty in our own existence needs to be dismantled. For this, Buddhist philosophy, with its doctrine of 'no-self', has proved a major inspiration. In Western philosophy, doubts about what Descartes rightly found most indubitable first started to seriously set in with Hume, who wrote that:

> For my part, when I enter most intimately into what I call myself, I always stumble on some particular perception or other, of heat or cold, light or shade, love or hatred, pain or pleasure. I never can catch myself at any time without a perception, and never can observe any thing but the perception.[15]

Whether or to what degree Buddhist philosophy influenced Hume has long been debated; it has even been argued that his wider philosophy, not just his soul-scepticism, bears this mark.[16] But in any case, his conclusion that our experience provides no reason to 'suppose ourselves possest of an invariable and uninterrupted existence thro' the whole course of our lives', was also that of the Buddha, who found that 'mind, intellect, consciousness, keeps up an incessant round by day and by night of perishing as one thing and springing up as another'. The 'uninstructed worldling' would do better to think of the body as the self, in fact, since it could last 'a hundred years, and even more'; although the ultimate truth was that there is no 'self' at all.[17] Hume just said we are inconstant 'bundles' of perceptions.

Richard Swinburne's reaction to Hume's Buddhist observation was:

> Hume says that he fails to find the common subject. One wonders what he supposed that the common subject would look like, and what he considered would count as its discovery. Was he looking for a common element in all his visual fields, or a background noise which never ceased? Is that the sort of thing he failed to find? Yet the self which he ought to have found in all his mental events is supposed to be the subject, not the object of perception. And finding it consists in being aware of different mental events as had by the same subject.[18]

Swinburne notes that it is among the 'basic data' of experience that successive and simultaneous experiences belong to the same subject, and that this is not something only known by inference, as if we were aware only of the object and had to infer the subject. Rather, we are aware of the subject in the manner, or way, in which we are aware of the object. The experience of a note in a melody sounds like part of the melody you are listening to. We do not infer from the experience that it must have the same subject that experienced the previous notes; we do not discover this as part of a later intellectual reconstruction. Rather, experiencing the connectedness of various experiences across time, and the unification of distinguishable experiences within a moment of consciousness, *is a way in which we experience a subject of perception*; just as experiencing greenness, shape and texture is a way in which we experience an object of perception, such as an apple. If I am eating an apple while looking at the moon, I experience a unity of taste and vision; the objects are not united, so I must be experiencing the subject. Experiencing connectedness and unification is not the only way, however, because we also experience agency. Some experiences feel active, we identify with them, and this too is a way of experiencing the subject. Having your arm moved is a different experience from moving your arm.

Kant brought the structuring, unificatory features of mind to the forefront of his philosophy, yet still agreed with Hume that we have no experience of the subject. 'The "I" is indeed in all thoughts', he said, 'but there is not in this representation the least trace of intuition'. For although we must represent an 'I' as accompanying all our thoughts, 'nothing further is represented than a transcendental subject of the thoughts = X. It is known only through the thoughts which are its predicates, and of it, apart from them, we cannot have any concept whatsoever'. Kant thought that to take this unknown 'X' for a substance was to fall for a special kind of illusion, which he named a 'Paralogism'. But if we do indeed experience subjects, in a manner appropriate to subjects as opposed to objects, then perhaps the idea of a subject as a substance – a soul – is in no worse shape than that of an object as a substance. Kant thought the concept of substance was 'empirically serviceable' only when applied to an object we experience as permanent, but as an idealist, did not think we could know the ultimate reality of objects.[19] But then, if the Humean intuition he fell in with is simply mistaken, since we do experience the permanence of the subjects that we are, then the concept of substance might be just as serviceable with subjects. If all consciousness is by its nature conscious of itself, that is, self-conscious, then it is hard to see how the intuition could not be mistaken. And it must be self-conscious, for if consciousness is to make one part of reality aware of another, it must be aware of itself bearing information about the other, which would be impossible if it were completely invisible to itself.[20]

From Hume (possibly), to Schopenhauer, to Parfit, the Buddhist doctrine of 'no-self' has provided inspiration to Western philosophers to deny that they really exist, in effect; and has provided many of them with consolation in the face of death to extol in their works too. These days neuroscientists have joined in; there has even been a volume published of some of them in conversation with the Dalai Lama, which includes the eliminative materialist philosopher Patricia Churchland.[21] What often tends to be neglected in this enthusiasm, it seems to me, is that Buddhism is a religion – as anyone who visits a Buddhist country will find it hard to miss – and that the doctrine of 'no-self' has an ethical purpose within that religion.[22] It is of course a metaphysical claim, and many great philosophers have argued for it. But exactly the same can be said of the doctrine of the immortality of the soul in Christian philosophy. Whether true or not, the doctrine of no-self finds a reason in the nature of reality for us to act in a selfless, generous way, and to be unconcerned if we do not get what we want in this life; just as the doctrine of the immortality of the soul finds a reason in reality for us to devote our lives to good deeds and the praise of God. A philosopher of either faith must find arguments to support these doctrinal commitments. A Platonist can say Plato was wrong about many things, but a Buddhist cannot say the Buddha was wrong about anything at all.

Behind this Buddhist doctrine is a powerful experience, which meditators achieve by sitting still for long periods, either looking at a blank wall or with eyes shut, depending on tradition, and 'trying' not to think or pay attention to any particular aspect of their experience. I say 'trying' in inverted commas, because while you are still doing *that* you have not succeeded. Galen Strawson describes the process as 'like maintaining one's balance on a parallel bar or a wire'. Experienced meditators can keep going for hours, even days. What they experience is described as 'selflessness', and it is unsurprising it should seem that way if our experience of self is of agency, succession and unification of diverse elements. The meditator is not moving, thinking or focusing, and there is nothing to distinguish one moment from the next. Yet the meditator is having an experience, one which somebody watching them is not; and the sense of well-being felt must be unified with other awareness. Strawson considers it a kind of 'radical' self-awareness: 'awareness on the part of the subject of experience of itself in the present moment of experience'.[23]

A more extreme experience of this kind can be produced by hallucinogens, which people have taken throughout our history. These produce extraordinary effects, often described as life-changing and profound, which can also be self-induced by very experienced meditators, or can spontaneously occur through sensory deprivation, devotional intoxication, fasting and the like. So-called 'ego death' is one of the most characteristic features. Timothy Leary, the notorious Harvard psychologist and LSD-guru, wrote (with others) a guidebook for the lengthy and multi-faceted experiential journey that modern hallucinogens produce based on the *Tibetan Book of the Dead*, interpreting this sacred Buddhist text not as a guide to what to expect and do in the period between death and reincarnation, which ostensibly is what it is, but rather as a guide to a distinctive kind of living experience which human beings are able to have, with 'death' understood as ego death, and 'rebirth' understood as returning to normal life as the experience subsides. Unsurprisingly, it was a book that caused a lot of bad trips.[24]

Ego death occurs early on, in the first of the three 'bardos', or transitional states, which the book describes; before hallucination and then 're-entry', as Leary puts it. Unlike in the gentle experience of ordinary meditation, the subject feels their ego forcibly falling away from them, in a manner they cannot control, to expose itself as an illusion. Attempts to resist this, by 'those who are heavily dependent on their ego games, and who dread giving up control', are ill-advised as a way of setting the scene prior to the hallucination stage, and similar problems can arise in the final stage if the person is in too much of a rush to get their ego back.[25] In such cases, Leary would have a guide say to the quite possibly terrified subject:

> You have still not understood what is happening. So far you have been searching for your past personality. Unable to find it, you may begin to feel that you will never be the same again. That you will come back a changed person. Saddened by this you will feel self-pity, You will attempt to find your ego, to regain control. So thinking, you will wander here and there, Ceaselessly and distractedly.[26]

The guide's advice, at this point and generally, is to 'let the stream carry you along'.[27] But what is it that the person feels they lose in ego death at the beginning, and struggles to reclaim at the end? It is the socially constructed ego, that is, the ego defined by our projects within the framework of ordinary, goal-directed life. It is something we freely create through our actions, and so, in the sense intended in Sartrean existentialism, we must continually choose ourselves. According to that philosophy, 'bad faith' occurs when we treat this constructed ego as a fixed entity from which our actions determinately flow, in a ploy to lessen the burden of our freedom. It is no wonder that someone engulfed in a tidal wave of overwhelming and unfathomable conscious experiences is no longer concerned by their usual goals, which define their constructed ego. The 'Game playing' of ordinary life, as Leary puts it, in which an ego is simply presupposed as we focus our attention on our goals, is forcibly put on hold when these goals can no longer be presupposed, with ego death the result. When the experience subsides, the person wants their ego back, but will not get it until they start living normally again.

The ego lost in this experience is what Dennett calls 'the self as a center of narrative gravity'.[28] Like a fictional character in a novel, it is the invisible point defined by all the actions of that character. But realising that there is something illusory about this does not mean we do not exist: that our selves are defined by our autobiographies in the same way evidently non-existent selves are defined by novels, as Dennett thinks. It would certainly be a big mistake to align such a view with the Buddhist doctrine of 'no-self', on the reasonable assumption that this doctrine received inspiration from the experience of ego death, since this would be to assume that the socially constructed ego is all we ever could be, such that if it is an illusion, we do not exist.[29] But the lesson of ego death, by stark contrast, is that we can shed these egos and learn from the experience, whether in transitioning to another life, in the supernatural version, or just by having a revelatory experience that teaches us to act in a less self-concerned manner; that is, by afterwards trying to construct a better self with the ongoing awareness that this is all it is, a construction. To take the lesson instead to be that you can *get what you want* – whether consolation in the face of the inevitability of death, or immortality for your socially constructed ego, such that it lives on as a computer simulation, for instance – strikes me as patently un-Buddhist.

The lesson to be learned from the manner in which Locke's theory of personal responsibility and the Buddhist doctrine of no-self have been distorted to meet the demands of materialism, I suggest, is that the characters our souls acquire as we go through life need to be carefully distinguished from our souls themselves. Try to make do with character alone, if your interests are metaphysical and hence exceed Locke's modest ones, and you risk allowing desires for immortality, consolation and control to drag you down the path of the grand *reductio ad absurdum*. Certainly we *care* about both character and soul when considering our own survival. Faced with the choice between absolute nothingness through the death of my soul, or the prospect of my character being wiped away for good so that my soul could continue with a new one, I would be indifferent. Without my character, I do not much care about my soul. But if the choice was between my soul continuing with my character, or coming to an end while my character continues by other means, whether in another soul or a machine, then I very much care. The former is life and the latter death, because my character without my soul is not me, only a simulation of me.

The end of my soul is the end of my conscious experience and hence the end of me. The end of my character would be just as bad, I have just claimed; but would it also be the end of me if my soul continued? I do not think it would, so I think the answer to the question of what we essentially are is simple: souls. The fact that we care just as much about our characters and souls, such that in thinking about personal survival we care only for their united continuation, not one without the other, is something that should not be allowed to confuse the issue.[30] So let us examine this imagined situation in which my character is completely erased while my soul lives on.

Looking forward to this prospect, what I face is just as bad as if my soul were to end too. But why is that? What would happen? The same conscious being would continue to exist but no longer think of itself as me. All the characteristics it used to exhibit, which allowed others to recognise that particular soul, and which formed the basis of its own self-conception, would be gone. And yet the same horizon of consciousness would be there. Reality would still open out on itself to be experienced from the same perspective, the perspective that used to be understood as belonging to something with a certain character, namely me. But what could 'I' have ever been apart from that self-aware reality? Certainly not just a social construction, since I am real. Rather, the social construction was a way of understanding my reality, one which altered it while it lasted, not essentially, but contingently. That same reality will continue without the construction, so I will continue, but I will no longer think of myself the same way. And looking forward to this prospect, the reason I do not care about this continuation is that everything I care about points to the construction. My hopes, dreams and loves are all referred to my reality, but only as characterized a certain way. It is my reality understood in a certain way that makes me want and value things, not my bare reality unaltered by this understanding. Without the understanding I do not care about myself because the caring is part of the understanding.

So, what is a soul? An experiential substance; that is, an experiential, enduring, individual existence. We cannot describe the ultimate, independent nature of souls any more than we can describe the ultimate nature of anything else, for ultimate reality is something we can only point to and make of it a minimal kind of metaphysical sense. Nevertheless, the concept of a substantial soul is 'empirically serviceable', in Kant's phrase, since each of us experiences a soul as the one most assured constant persisting throughout our lives. Perhaps an individual conscious existence could not continue unless the objective sense we make of reality contains a continuing human body; perhaps only a part of that body must continue, such as its brain; but certainly it could not continue unless my experiences remain unified in the moment, unified across time, and I continue to identify with at least some of those experiences as my own activity. The soul is what unifies and has those experiences.

The terminology of 'substance' is an ancient and vexed one, but there is a clear philosophical point to it which remains invaluable. As Tim Crane has argued:

> the fundamental motivation behind a belief in substance is the idea that there are kinds of particular entity in the world whose existence and unity is not simply a consequence of stipulation, postulation or theory-building. Their unity is rather explained in terms of some underlying natural principle which is responsible for their organisation, activity and boundaries.[31]

Understood this way, to claim that something is a substance is simply to claim that it is essentially a bearer and unifier of properties, such that without those properties, it is nothing at all; hence the persistent concern that, in itself, a substance must be a curious 'featureless substratum' is clearly a confusion, as Anscombe once pointed out.[32] A soul is a substance because it is essentially a bearer and unifier of experiential properties. To commit to substances need not be to make a claim about ultimate reality, so we need not follow Locke in thinking of substance as unknowable. An idealist need only claim that the concept of substance is integral to both objective and subjective thought, since it is needed to distinguish unities we create through our classificatory practices, from those we feel reality itself must be responsible for.

Crane argues that people are mental substances, that is, that 'the principle of unity of a person is mental', because the unification of a mental life around a single point of view both at a given time, and across time, is what constitutes the unification of a person.[33] Since he rejects the traditional criterion of substance as independent existence, largely on the grounds of vacuity, this allows him to claim that:

> if persons are substances, the possibility of disembodied existence does not show that a person could exist without their body; all it shows is that *something* mental could exist without a body[34]

So although a substance is essentially a bearer of properties, and the defining properties of a mental substance are of course mental properties, to claim that a person is a mental substance still leaves open the possibility that the substance could not continue to exist without continuing to unify non-mental, bodily properties. Upon disembodiment, the mental substance that is a person might cease to exist and be replaced by another mental substance which is not. In that case, bodily properties would be essential to the existence of a person. But Crane is not claiming that only mental properties are essential to the existence of the person, since he rejects the criterion of independent existence for substances; rather he is claiming that the unity provided by the substance is a mental unity, a unity which might govern non-mental properties essential to the existence of that substance.

Although I think this is a good proposal for how to use the term 'person' for certain purposes, just as Locke's was, both distance us from the question of what we essentially are. For if it were possible for my conscious experience to continue past the demise of my body, and I found myself in a situation of disembodied existence, then if I accepted Crane's understanding of 'person' I would have to conclude that I was no longer a person. But in that circumstance, this does not seem significantly different from the obvious realisation that I am no longer a man, and the further conclusion I would have to draw is that I never was essentially a person; I cannot have been if I am still here but no longer a person. Of course, I am imagining that my personality would continue, my socially constructed ego. If it did not, then such a future is no better a prospect for me than death; it is the prospect of becoming a mere '*something mental*', as Crane puts it. And yet if my soul did continue to unify a mental life around a single point of view, albeit not in a manner that preserves my personality, then the substance that once bore my personality would endure. And that substance is all I can have ever essentially been, given that I am an individual and not a type. I would still be there, but no longer understanding myself in the manner I care about, I might as well not be.

Although the very idea of a horizon of consciousness precludes us from conceiving it as a substance in the sense of an independent existence which appears to us – a horizon is rather *that within which* things appear – on Crane's understanding of a substance as a bearer and unifier of properties, there is no obstacle to thinking of a horizon as a substance. In fact, a horizon then becomes an exemplar of a substance: it unifies everything that appears within it as *my* experiences, conceived subjectively, and as appearances of objects to me, rather than the objects themselves, conceived objectively. When it ceases to bear any property of having something appear within it, then it is nothing at all: there is no 'it'.[35] This latter prospect, of my horizon permanently closing, is the prospect of my death, and it is for this reason that Valberg says the horizon 'assumes, as one of its guises, the guise of the self'. In addition to this horizonal conception of self, he also explains that we have a 'positional' conception: I am the human being at the centre of my horizon, where 'at the centre' is explained phenomenologically to capture the sense of what I called the 'body-subject' in Chapter 6.$_{36}$ But the positional conception is premised on the horizonal one, and it is the demise of my horizon that constitutes my demise. So, the essence of Valberg's position is preserved, I think, by thinking of horizons as substances in Crane's sense. And this allows us to simply say: we are essentially souls.

§3. Video Games

On the jazz album *Song X*, Ornette Coleman played his composition 'Video Games' with Pat Metheny and others. Both the melody, and to a greater extent the improvisations by Coleman and Metheny, capture in music the essence of the old arcade games of the 1970s and 1980s. Although there is direct imitation of the sounds, most notably via Metheny's guitar synthesizer, what is most uncanny about the representation is how they capture the experience: the excitement, the total engagement, the tense, precarious progression interrupted by temporary respites, either as you move up to the next level with glee, or mess up and have to start over again. Video games have changed a lot since then.

These days they are more like real life. Or at least some are; some still follow the original model, some are like novels in which you control the protagonist, and there are many other approaches to the essential idea of playing a game by interacting with a virtual environment. The games that interest me, however, are those people play for years, may well expect to play for the rest of their lives, and in the case of the most dedicated players, commit to at the level of a professional sportsperson or musician, receiving similar levels of adoration within their online communities. They are not like real life in the obvious sense that nobody is a Samurai warrior anymore, nobody has ever been an interplanetary space trader, and however much real football managers and airline pilots might marvel at the verisimilitude, the players are still only looking at a screen while operating a controller. Rather, they are like real life in the deeper sense that they create ongoing tasks, which contribute to wider plans, which in turn contribute to still wider plans and ultimate ambitions, with these plans often moderated by realism as you 'play' and learn. Each such game is an artificially created social framework. Quite unlike the old video games, and just like real life, much of what you do is boring; 'grind', gamers call it. But those endless hours of boredom pay off if they allow you to make *progress*, by increasing the powers of your character, or by earning the money needed for better equipment, say. Weeks or months of frustration and occasional despair can be worth it, since there is no other way to achieve your aims – and there are bound to be small successes and pleasures to keep you going. Just like life, having fun may become a guilty pleasure in a game, since it is bound to involve doing something you can already do well, when you really ought to be working on that difficult skill – work you keep putting off, but promise yourself you will get around to eventually, although perhaps never do.

I suspect many people of middle age and above have practically no idea what some games are like these days. Second life is no longer an idea, but a mass reality which will grow and grow. Before long, when true realization dawns about the immense addictive powers of an alternative life in which your role in society is much more exciting than what it actually is, I expect we will see governments adverts for real life popping up to annoy gamers; glossy reminders of the pleasures of sex and alcohol, perhaps.[37] Here is some testimony from one of my postgraduate students:

> One year during the peak of my WoW (*World of Warcraft*) experience, I put in over 2,500 hours. So, 105(ish) days in 365. I would go to school, get home, play WoW, eat, play WoW, play WoW, play WoW, and sleep at 4am, 3 hours before getting up for school. For my real life, this wasn't good, clearly. But in-game? I had almost the top tier of raiding gear, I had a pretty rare red proto-drake to fly [...] Do I regret it now? Not really. And I think this is the point that's stuck with me. When I think back to my memories of playing WoW, they feel like actual memories [...] I remember the feeling I got when I entered Stormwind City for the first time, and I was completely immersed in the sheer scale and atmosphere. I had no idea anything could feel so big and real [...] these in-game memories, both in vividness and in emotional force, have pretty much the same feel as special memories I have of the 'real' world.[38]

Some young people now dream of being 'pro gamers', which is a real option; gaming houses may pay accommodation and living expenses for players following a strict and intensive practice regime, and with sponsorship, crowdfunding and tournament prize money, a new kind of celebrity lifestyle has become possible. But which future has more hold over their dreams: what they will be and do in the virtual world, or the real life that makes this possible?

The idea that there is a meaning of life is, essentially, the idea that our creator has a masterplan for the human race. Video game developers ('devs') create virtual worlds and maintain their masterplan (God is devs). Real life can never be a game because games are our invention and human life is not, but virtual life is indeed a game, and if it is what we increasingly do with real life, we approximate a meaning of life as best we can. Unlike a god we can never be certain of, however, game developers are a known presence. The power they wield already causes great resentment when players feel unreasonable demands are being made of them, or when they dislike the direction the game is being taken, whether through direct action, or by the influence of certain players being left unchecked; this becomes especially acute when real monetary interests affect virtual life, as often happens – real life is seen to defile the virtual. But if the attractiveness of virtual reality increases faster than the attractiveness of real life, the devs may start wanting to invest the capital of their godlike powers not for the sake of the game, as a true god, nor for the sake of their advantage outside the game, as a corrupt god, but rather for the sake of their own character within the game. Human maintenance of virtual reality may become unsustainable for reasons of equity. Then we will truly be in the business of trying to make a god.

This new direction we have just started taking is rooted in the soul's natural tendency to solipsism. Many video games are played with others, but still the gamer sits on their own, or, so far as the game is concerned, might as well – the vast time spent on them can make this practically obligatory. Some games offer the choice between interacting with others or only with AI; useful if you just want to get on with something without relatively unpredictable people getting in the way. But the virtual others you interact with, even when controlled by real individuals, are, from your perspective, just a *type* that might be replicated perfectly by AI. Similarly, in real life, the people we interact with could in principle be replaced by robots without our noticing the difference, which illustrates the traditional 'problem of other minds'. But if the faces we look into and hands we touch did not have a soul behind them, the deception would be massive; which is not so obviously the case with the game parallel. Certainly, the fact that the pixels change on the screen as the result of a real person operating their controller is known and responded to accordingly by gamers. Most would find it far more galling to have their spaceship destroyed by a real person than an AI, for instance, because then they lost to a person, and you cannot really lose to a machine, only not play well enough. But note that this is a response to real life; within the game, it makes no difference. In virtual life, considered in the ideal purity that captures our imaginations, there is no difference. So, if the pull of real life dissipates, as seems to already be happening, and it ceases to be our primary source of evaluation, then we may well start to *feel* that it makes no difference. Our connection with others may become a tenuous intellectual construct, rather as, at present, scepticism about other people's souls is a tenuous intellectual construct.

In 1974, Robert Nozick imagined an 'experience machine' which we could plug our brains into to enjoy a life of maximally rewarding experiences.[39] He found three reasons why we should not. The first is that 'we want to *do* certain things, and not just have the experience of doing them', the second is that 'we want to *be* a certain way', and the third is that the experience machine 'limits us to a man-made reality'; this third reason, curiously, he sees as clarifying 'the intensity of the conflict over psychoactive drugs', because some regard them as like an experience machine, while others think they reveal deeper levels of reality, and hence show us a good reason not to plug in. But in video games, unlike the experience machine, the experiences are not just passively received, which negates Nozick's first two reasons. The gamer does *do* certain things and they *are* a certain way in the game. Games can lead people to neglect what they do and are outside the game, of course, but Nozick's only reason for thinking real life superior is provided by his third reason. And here, the idea that a man-made reality limits us, in a manner Nozick is exceedingly vague about, is easily countered with the clear thought that it is a reality where limits, such as the body you were born with and society you were born into, can be overcome by human ingenuity.

What I find odd is that it never seems to occur to Nozick that plugging into such a machine would cut you off from other people. Pining for the people you left behind would interfere with the bliss the machine generates, so it would have to negate it, if indeed this would even be necessary; as I doubt, given what real narcotic addiction is like. The only good reason not to plug in, so far as I can see, is provided by human relationships, especially those based on love. Video games need not cut us off from others – they provide a new way of interacting, which can bring people together and provide cherished shared experiences – but they certainly have this tendency, rooted in the nature of our souls.

In a powerful piece of philosophical analysis, Valberg asks why the prospect of death can strike us as unfathomable and impossible.[40] The analysis draws on an episode from Tolstoy's *The Death of Ivan Ilyich*. Ivan sustains an injury, thinks little of it at the time, but it gets worse, and so he puts considerable effort into getting it medically treated. Eventually it dawns on him that he is going to die. He 'did not and could not grasp it', says Tolstoy. Ivan remembers the syllogism: 'Caius is a man, all men are mortal, therefore Caius is mortal' – this 'had seemed to him all his life to be true as applied to Caius but certainly not as regards himself'. In his despair and bewilderment, he says to himself: 'it can't be, and yet it is!'

I cannot do justice to Valberg's analysis here, especially since it connects with the whole of his philosophy, but I can state the conclusion it suggests to me. The reason death seems impossible to Ivan is that the soul has a natural tendency to solipsism. For Ivan, the whole world is something which appears within *his* horizon of consciousness, so if that horizon closes, the world will cease to exist. And yet he knows it will not; hence the apparent impossibility. This tendency is unsurprising for a self-conscious being whose consciousness of its own experiences provides its consciousness of the rest of reality. Hence we should be very wary of using technology to accentuate and reinforce this tendency, since it is directly opposed to the altruistic needs of morality.

I think virtual reality has the potential to provide us with an exceptionally fulfilling future, precisely because it is designed to satisfy our wildest dreams; and as we live them, it will inspire new ones. But we are not remotely ready. Without widespread reflection on what we should do with it, in which philosophy could play a vital role, all our old problems are likely to be reproduced. And by turning to virtual reality, we will turn away from real world problems, while weakening ourselves against those who would take the world in directions we do not want. Once we have sought the help of the titans, we never forget what they taught us, so we will always have video games, and the people playing them now probably always will. But a new technological innovation which allows considerably greater immersion, such as by allowing direct interface with virtual reality without the need for screen and controller, is something we should be trying our hardest to prevent. Instead we are racing to that outcome, which is already here in primitive form. In a peaceful, equitable, philosophical world, in which we took good care of other animals and the environment, I would like to see video games made more soulful, in the sense that music can be: you put how you feel out into the open, for all to see, and thereby connect with other people who feel the same. In our world, universal philosophical education would be my priority. If the consequences of thinking of ourselves as types were thought through, then we might think of ourselves, and the collective directions we are currently taking by technological means, very differently indeed.

8

Truth

§1. The Value of Balance and Post-Truth Culture

We have to strike a fine balance in our personal dealings with the truth. An unyielding reverence can result in insensitivity, while too little respect can break down trust and stifle rationality. Harry Frankfurt, in an essay about one form of the latter imbalance ('bullshit', he calls it) provides a good example of the former.[1] He recounts an anecdote in which Wittgenstein visited an acquaintance who was recovering from having her tonsils removed. 'I feel just like a dog that has been run over', she croaked. Wittgenstein was 'disgusted' on the grounds that she did not know how a dog in that situation would feel; according to Frankfurt's analysis, what angered him was her disregard for truth. Somebody with too much reverence for truth might tell a child their painting is terrible or a colleague that their new suit is unflattering. Whether too little respect for truth is shown by telling the child it is 'brilliant' or the colleague that they 'look great' is a moot point; the comment will not be constructive, but then it is doubtful that constructive comment was called for. Nevertheless, if somebody always just says what they think you want to hear, or what it suits them for you to think that they think, or what it suits them to have you to know they said, whether or not they expect you to be taken in by it, then you are left guessing about what they really think, and the possibility of rational discussion recedes. For such discussion to proceed, you must first get their actual view into the open, which, if their disregard for truth is an ingrained enough habit, may be hard to do; it may be a matter of getting them to work out what they think for the first time, and might not even be possible. Striking the right balance between too much and too little regard for truth requires situational awareness and is an ongoing problem in developing our characters, or should be.

In public life, we have had considerably less success in striking this balance than most do in their private lives, and it is abundantly clear on which side we have fallen. We do not have a problem with our politicians being too brutally honest. We do not have a problem with companies being unable to sell their products because they admit their faults. We have created an atmosphere in which even a known truth sometimes cannot be spoken. This phenomenon is most notorious in totalitarian states, where to speak the known truth may be to sign your own death warrant; when nobody can say what went wrong with the nuclear reactor, because it contradicts the official line, then nobody can take the action required to prevent a repeat occurrence. But the same stultifying and repressive effects can occur whenever there is insufficient respect for truth. The following example is of a kind which ought to be familiar to everyone these days. A politician says they will do X by date Y and absolutely no later. A journalist then points out various foreseeable circumstances in which the politician would obviously delay doing X. The politician knows this is true, but will not say it, and for good reason: the press would lead on it, the public desperate for X to happen would be infuriated and a rival would take the advantage. Why did the politician make the promise? Because that is what a significant proportion of the public wants to hear. They want to hear it whether or not it is realistic, because they lack respect for and interest in truth. So, they get politicians to match.

We are said to have entered a 'post-truth' era in recent years, in which factual claims are made primarily for their emotional force in advancing some agenda or another, with factual rebuttals then ignored, or else disparaged on the grounds that they come from 'experts', who are not to be trusted because their apparent respect for truth is bound to just be a front concealing an alternative agenda. If the truth becomes too conspicuous, and hence interferes with the emotional force required (typically anger or fear), then you move onto the next claim, which our technology of instant mass communication makes it easy to do. When the truth rears its ugly head you just change the subject. The appearance of truth is a useful tool for feeding the wave of emotion, and the actual truth, if indeed there is such a thing, is neither here nor there. Everyone who recognises this phenomenon, and thus names it 'post-truth' this or that, strongly disapproves.[2] And yet you could be forgiven for thinking that society is just assimilating one of the main lessons about truth taught by twentieth century philosophy.

§2. Philosophical Scepticism about Truth

The scene was set by Nietzsche, who thought that 'Truth is the kind of error without which a certain species of life could not live', and that knowledge, which presupposes truth, is just 'a measuring of earlier and later errors by one another'.[3] The idea that there is a truth about reality which is not just useful error, according to Nietzsche, was invented for the sake of advantage; a naturally inferior class of people used 'the truth' to dictate how everybody should live and gain advantage for themselves. As a remedy, he advocated a perspectivism in which we recognise that truth is always a function of our personal will to power, that is, of the errors and illusions that best serve our own purposes. This anti-authoritarian line of thought became a dominant strand of twentieth century continental philosophy, probably the best known one, for which Derrida provided the slogan: 'There is nothing outside of the text'.[4] There are no binding truths, only negotiable interpretations. The postmodernist philosopher John Caputo (one of the few happy to endorse the 'p' label) is quite clear:

> I am saying that the moment someone says this *is* the truth – this *is* democracy or science, sexuality or ethics – the *one true* interpretation, then the flow of truth is cut off, the borders are closed, the event is prevented, the life of interpretation is crushed, the future is shut down and replaced by anxiety about the future. The police of truth have arrived.[5]

Since society at large can hardly be expected to grasp all the subtleties of philosophical discourse when philosophy is barely taught, a post-truth culture seems a reasonable enough response to this kind of message.[6]

The founders of American pragmatism came to much the same conclusion, as encapsulated in William James' statement that 'The true is the name of whatever proves itself to be good in the way of belief'.[7] The mainstream of twentieth century analytic philosophy, on the other hand, gravitated towards a kind of deflationism, inspired by Ramsey's view that truth is redundant – since to say 'the cat is on the mat' is true, is just to say that the cat is on the mat – as well as by Tarski's theory of how to define truth for any given language in terms of conditionals of the form: 'the cat is on the mat' is true if, and only if, the cat is on the mat. The latter led to various subtle views according to which truth is not in fact redundant, and is perhaps very important,[8] but nevertheless, if anyone outside of academic philosophy did happen to catch wind of any of this, the headlines are hardly of a kind to inspire greater respect for truth.

Rorty brought all three strands together – continental, pragmatist and analytic – in a clear and easily digestible form.[9] Nietzsche was right that there is no truth about reality; the whole idea is just a God-substitute, an authority we invented because we were too weak to face the fact that what we believe is our responsibility and ours alone. The original pragmatists were basically right, but should not have tried to define truth in terms of usefulness to humans, since it is rather *justification* which is usefulness to humans. Nevertheless, the only criterion of truth we can ever have is justification, so the resulting 'neo-pragmatism', as some call Rorty's position, amounts only to a technical refinement of the original. And when it comes to truth itself, the non-pernicious kind appealed to when we say 'the cat is on the mat' is true, but without trying to back this up by appealing to reality itself, the deflationism of the analytic philosophers is essentially correct. We use truth to commend what people say and, sometimes, to urge a little caution; such as we might say that a scientific theory may be justified now, but still not true – all we really mean is that a new, improved scientific community of the future might not consider it justified. So, truth is a terrible, repressive idea if misconstrued, and a useful linguistic device if correctly construed. We should have no respect for truth, only for justification to our various communities.

Rorty makes justification seem like a barrier to reality; one so impenetrable that doubts about whether there is anything on the other side might even start to seem reasonable. He reasons as follows. If a theory is considered thoroughly justified within the expert community best-suited to judge it, then it is taken to be an established truth. But does that mean it corresponds to the nature of reality? To know that it did we would have to 'step outside of our skins', as he memorably put it, to compare reality itself with the description humans consider most justified – then we would be able to see how well the two matched up.$_{10}$ But we have no way of knowing reality, and hence knowing the proper way to describe it, apart from the methods which produced the description humans consider most justified. So since we have no way of comparing that description to reality – as opposed to comparing it to reality as presented in the description, to which the description will obviously match up perfectly (from here, the deflated linguistic truth-device falls out trivially) – it seems to follow that if adequacy of the description to reality is what truth amounts to, then truth is an illusionary aspiration. Justification of the description to other humans is as far as we could ever go.

Rorty thought this was a pragmatic proposal: a useful, practical way of thinking fit for a liberal, democratic society. But its origin in Nietzsche's maximally illiberal idea of conversation as a grab for power was not tamed. For if the only criterion of truth is justification, and a justified belief is just a useful one, as Rorty thought, then somebody must decide what is useful – as nobody could decide what is true, except in a fiction they create. If that decision concerns collective action, where there is bound to be disagreement about what is most useful, then the decision will be made by those in power, just as it has always been. And without the authority of truth to bind them, other forms of justification start to seem more legitimate. Justification in terms of the truth was only ever *one* kind of justification. You can also justify your description of reality with a lie, or by the anger or fear your description creates.

To provide a justification, you give your audience a reason which justifies it to them, one which persuades them. Anger or fear can be such a reason, as can a lie – a lie which you took to be truth can still be *your* reason. But a liar is not trying to help you reach a true conclusion. And if somebody gets you angry by telling you about something they think happened, but are not really sure and do not much care, then they are not building a case, a justification, designed to help you recognise the truth. All either cares about is what their claims will make you think and do, and how this might satisfy their own desires. Neither the liar nor the disregarder feels constrained to offer a justification motivated by, and made in terms of, truth. Liberated from that idea, they think only of usefulness. The liar thinks only of usefulness when they will into existence a useful error (as Nietzsche might have put it). The disregarder, the authentic denizen of post-truth society, also thinks only of usefulness, and will use any tool at their disposal to achieve it, whatever works best (now it sounds more like Rorty). Both think it will be useful if you believe them – useful for them, that is. And it might be useful for you too, depending on how nice and clever they are.

You might reasonably object that neither is providing a real justification; that if you are lied to, your view is not justified, you just think it is. This move would require an appearance/reality distinction made either by reference to truth or a privileged kind of justification. On the simplest version of the former, only a true belief can be justified. But innumerable false beliefs were held in the past which were justified in their time. So, the link will need to be made in terms of what the justification is supposed to do – either direct you to the truth, or persuade you in another way. Then we could say that since the liar does not try to direct you to the truth, their claims cannot provide justification; and that the honest inquirers of the past had justifications which failed to direct them to the truth. Personally, it strikes me as simpler, and less taxing on our concept of justification, to just say that a view is justified if a justification is accepted, but that not all justifications reveal, or aim at revealing, the truth. But either way of talking is fine; the value of truth to justification is just being emphasised in different ways.

The other way of trying to rule out views being justified by lies, or just to disparage this kind of practice, would be to have an appearance/reality distinction based on a privileged kind of justification, so that truth need not be mentioned. Then we might say that the person who is lied to does not have a (proper) justified view because lying falls short of good justification. If Rorty were to take this option, and he did come close, he could then say that certain practices of justification are better than others because they are more useful than others.[11] And yet lies can be exceptionally useful. Not for the collective, you might suppose, not in the long-run. But they might be. If the leader is benign, the lie that justifies a collective action might be the most useful option for everyone. If the populace later discovered the lie, they might agree, seeing that without it they would never have been persuaded to take a course of action which greatly benefited them. So, it is hard to see how a liar or disregarder could possibly be excluded from having the ability to provide good justifications without mentioning truth. To appeal to openness, integrity and diligence as signs of good justification is just to piggyback on traditional indicators of justification that aims at truth. It is to forget about the idea of justification as usefulness to disparage dishonesty on the grounds that the truth will out.

If Rorty promoted a view with foreseeably bad consequences, he could hardly retort: 'I can't help that ... what I'm saying is still *true*.' On his criterion, the justification of a view is its usefulness. But then, what makes something useful? It cannot be the truth about reality – that the weight, size and shape of the hammer make it useful for knocking in nails, for instance. These are just useful descriptions. But that means the hammer is useful because it is useful to think it is ... and if that just leaves you wanting to know why it is useful to think it is, the answer can only be: because it is useful. What this shows, quite clearly I think, is that usefulness is a useless substitute for truth. It is an evaluative concept, and you cannot evaluate unless you have something to evaluate.

§3. Truth First

Reality and truth must come first. The alternative is incoherence, which you can only live with through anti-philosophy. The way in which Rorty and the traditions he represented made justification seem like a barrier to truth, when it is really our method for getting at truths that have not revealed themselves, was by turning it into their own truth. They extracted the image of people exchanging reasons from the world of truth, then rode it to an imaginary place where disembodied language is endlessly interpreted by the reader who forgot herself, or where usefulness relates to usefulness in a self-sustaining mesh of nothingness.[12] The most immediate historical root of this confusion seems to have been thinking that Darwin's science of evolution had great philosophical significance for how we think about truth; this influence is well-known for both Nietzsche and pragmatism. We are invited to imagine all those animals running around making mistakes that aid their survival – they think these mistakes are the truth, but only because they proved useful. But that only makes sense if we suppose that *we* know the truth, and hence can recognise not only the errors of the animals, but why those errors are useful to them. Since we number among the animals, then, it makes no sense at all: it requires us to do exactly what it is supposed to motivate us not to, namely 'step outside our own skins'.

We need not worry about being confined within our skins, however, since we are self-conscious realities. As such, our existence is a revealing of truth. Thus revealed, truth is what we tell, not a relation between the telling and the told about – between our descriptions and reality. The latter idea can only take hold once a world of revealed truth has already been presupposed, one in which there are such things as descriptions and relations, but not – if we find ourselves led down this route – the suddenly estranged reality which can never be revealed. To tell the truth is just to describe reality that has already revealed itself, with justification coming into the picture only when our descriptions are disputed – which presupposes shared knowledge of the truth described – or when truth has not been revealed. For example, I can see a rose in my garden and I feel a little tired because I stayed up late last night; these truths have already been revealed to me and I just told you them. The question of justification only arises if you dispute my descriptions – as nobody ever would in this case – or if you doubt I was telling the truth.

Any attempt to get behind truth, to try to understand its genesis from something more basic, will always presuppose its own truths – such as animals interacting with each other in useful ways – and if it is insisted that this too is only interpretation, then either interpretation is reality, or reality does not exist, both options being absurd. This kind of thinking results from letting the idea of 'interpretation' run away with itself. We *notice* things in everyday life – like something moving in that tree – and the question of interpretation arises when they are unfamiliar or immediately ambiguous, and hence require some interpretation. We look for an interpretation not to notice the truth we already noticed, or to turn reality into a truth, but just to find the right way of describing it so we can convey it to another, or follow through on some implications of the description we used. Only if you think in the metaphysical context of the ultimate nature of reality does it make sense to think of all our most conventional descriptions – those which convey truths in everyday life without the need for active interpretation – as themselves a kind of automatic interpretation. But to think that way is to lean on the ordinary notion. When the explicit context of discussion is ultimate reality, then 'it's a bird', or, 'I saw a dark shape moving in that tree' can be *thought of* as interpretations, but only by relying upon a notion of interpretation which takes it for granted that ultimate reality – not described as such, or indeed in any way – reveals itself in truths we can describe by talking about objects and experiences. If you make everything seem unfamiliar, as metaphysics sometimes must, then the question of interpretation will of course become prominent. But truth is at the bedrock of our thinking and interpretation is not.

Justification and interpretation guide our descriptions of the truth, and can generate truths of their own, but those who live among the truth do not always need a guide. Williamson has argued that since attempts to analyse knowledge in terms of justification have consistently failed, and seem to do so in principle, we should think of knowledge not as a kind of belief – the justified kind that aims at truth – but instead as a different kind of mental state: one which puts us directly in touch with the truth.[13] Perhaps that could provide a good way of thinking of our life among truth: as an ever-present state of mind. This need have nothing to do with language, as I would understand it. A dog cannot tell the truth, but reality still reveals itself to it, and it behaves in accordance with those truths in its own way. How it would reveal itself to a dog that has just been run over we can only guess at, as Wittgenstein immediately realised, but it cannot be right to think there is only a truth for us language-users, not the dog. If reality revealed a truth about itself, it was to the dog.

Reality must have a truth about it, because this is a prerequisite of coherent thought. Deny it, and you deny any correct interpretation of the thought itself, so the denial undermines itself. But that does not mean we can express the truth adequately. We do our best through a combination of the framework and objective and subjective thought, but metaphysical reflection reveals inadequacies once we start thinking about how these interpretations fit together, or more obviously still, how the totality being interpreted came about. When I say 'there is a rose in my garden' or 'I am having an experience of a rose', I tag much the same easily communicable feature from the vast wave of truth revealed in that moment, just with different implications; the latter tells us reality would not have revealed itself thus if I had been lying unconscious, for instance, while the former suggests many other ways reality might have revealed itself in that moment, to me or to others. Neither was adequate to the truth revealed in that moment, but I picked up on some of it in an uncontroversial way and thereby told the truth.

Only because reality has revealed the truth about itself can we use reason to try to get at other truths that are not revealed, whether just to ourselves, or to anyone. We reason about truth that might yet be revealed, and about how things must be given the truth that is revealed. The former is a daily preoccupation, as we reason about why somebody did what they did, or what caused that noise, for instance. The latter is a more theoretical matter, in which truth becomes a demand of reason. Thus, we theorise about why Caesar crossed the Rubicon and what the fundamental constituents of matter are, for instance. The only truth that reveals itself in these matters is the evidence, such as what Caesar and his contemporaries wrote, and what scientists have observed in their experiments. These are not cases of describing revealed truth as best we can, but rather of trying to follow through on its implications with reason. In the history case, we may suppose that a truth did once reveal itself which would settle the matter – what Caesar was thinking – but in the science case there is not even that.

§4. Philosophy's Traditional Obligation

Truth is of the utmost importance to our world because people continually make decisions about which truths to reveal and which to conceal. We cannot reveal all because we are temporal, and we must choose because we are free, so there is a balance to be struck. Since concealing a truth can give you power over others, and is so easy to do for souls, we live in a world full of deceit, where honesty is one of our most highly regarded virtues. For this most obvious of reasons, a philosophical assault on truth was never going to be helpful in our ongoing efforts to maintain this balance. Even if the reasoning had been good, its conclusion about the unimportance of truth could only ever have applied to a theoretical notion of truth, not the one our world hinges upon.

At one extreme of the balance we must strike, imagine a world where all our truths – everything revealed during our lives, including our thoughts – are publicly-accessible knowledge. This is a direction we have set out on by technological means, as surveillance of all kinds increases.[14] You do not need to have anything in particular to hide to find that prospect hideous; some things are best kept private, and it is because they are that we can be continually surprised and interested by the world, with some of it remaining special to us. Complete openness is not suitable for humans and it is hard to imagine how even gods could put up with it. Then, at the other extreme, imagine a world where we keep all our truths to ourselves. This is another direction we have set out on by technological means, as we saw in the discussion of video games at the end of Chapter 7. Without the gods to metaphysically balance our possession of both objective and subjective thought, the titans will lead us in both directions at once. We will cut ourselves off from those we care about, while others watch us.

Respect for truth is of particular importance to public life because it is required for collective rationality. It is only against a shared backdrop of truth that people can come to rational decisions about what to do and think. Without that, whatever justification sways one person cannot be assessed by the other, so if there is something I cannot tell you which persuades me of X, then, as a rule of thumb at least, I should not be trying to persuade you of X by other means. Collective rationality requires as near to a shared route from truth to decision as is practically possible. And that requires a collective interest in truth and reason. Philosophy has a traditional obligation to promote that interest. So: how well has it been doing lately?

Dennett thinks that 'what the postmodernists did was truly evil', because they 'made it respectable to be cynical about truth and facts'.[15] I agree that 'postmodern', or better, post-Nietzschean philosophy, has been particularly unhelpful in discharging philosophy's traditional obligation. The mainstream of twentieth century philosophy was materialism, however, and it remains the most influential philosophy of our day. So, what has materialism done to discharge philosophy's traditional duty? It has cast doubt on subjective thought, by raising the possibility that our conscious experiences are illusions, and even that we ourselves are illusions; for to say that the self is an illusion is just a misleading, distancing way of saying that we are ourselves are. It has similarly cast doubt on objective thought, by making us suspect, for example, that the rose in my garden might be in need of vindication from scientific theories about fundamental matter. The suspicion of illusion which materialism continually raises is itself in jeopardy, in fact, because if illusions are false judgements, states of machines, then they might also need vindication from scientific theories of matter. Illusions might be illusions too. We have embraced a philosophy which casts doubt on all of the truth that reveals itself to us in our ordinary lives – all without remainder.

By following through on the demands of their metaphysic, materialists have tried to replace revealed truth with theoretical truth. To do so required a covert extraction from the world of truth, just as with post-Nietzschean philosophy, since something must remain unquestioned for theory to proceed. And again, justification and interpretation was chosen – but this time only that of science. The result has been a widespread feeling that the truth about reality is not the ordinary person's business: that to find out about it, you must look up the latest results you will not understand except through highly metaphorical language produced by someone who does. Those who make the effort will then find it confounded when news of the latest scientific breakthroughs come in. Confidence in individual reasoning has been undermined, as we are told about the natural biases that only those employing scientific methodology can hope to avoid. There is the constant suggestion that none of us really knows their own mind, what they themselves think, because this can only be determined by testing us in large groups, or by scanning our insides. Many have learned to think that an alien agent called 'the brain' is doing the thinking for us anyway, in accordance with its possibly dire evolutionary imperatives, themselves set by the physical imperatives of the most ancient and incomprehensible event. Against this backdrop, which might be grasped in no more cogent form than 'science has all the answers', it makes sense to forget about truth and concentrate on the errors you know in your everyday life. With justification in terms of truth unavailable, respect for it becomes a kind of piety to science, rather than a practical imperative. And for those without that piety, post-truth society offers an alternative.

Technology fundamentally changes how reality can reveal itself as truth. When the atomic mushroom cloud first rose up, reality revealed itself in an unprecedented manner, and the power to make it do so again was forever placed in our hands. Despite this strongest possible warning, however, we have not learned the importance of striking a sensible balance with scientific truth. The balance to be achieved is not between what to reveal and what to conceal, which is a problem for us all in both private and public life, and one which we do put effort into, with mixed results. The balance is between what to reveal and what to leave concealed. This is more difficult since we are dealing with unknowns. But every investigation has direction and aim, determined by what is already known, so the unknowns are only relative unknowns. We could certainly try to strike a more cautious balance. Instead the complete lack of balance is palpable: competing sides try to discover everything they can as quickly as they can. As far as public life is concerned, this is our essential philosophy and politics of science and nobody is even trying to hide it. Public consciousness of the need for balance in this all-important area has barely risen above pointless clashes between pessimism and optimism, with predictions of the future based on today's *status quo* attaining far more prominence than plans to safeguard our future. Here is a question of balance which is conspicuously absent from our public life.

To discover everything there is to know about the physical universe, for materialists, would be to know everything there is to know about ultimate reality – to achieve omniscience, to which omnipotence is the traditional accompaniment. Is that what is driving us? I suspect more prosaic motivations, un-reflected and contradictory, with the main role played by materialism being the promotion of anti-philosophy. Materialism is an obstacle to asking how much we want science to tell us, since it recognises only what science does tell us. It can only be disputed with philosophy, yet asks only for reasoning based on experimental results. But materialism was not established by experimental results and will never be falsified by any either; it has that much in common with the vast majority of things we believe. The same can be said of the positivist's rejection of metaphysics. Surveying people to acquire an experimental result does not bypass the need to ask why each of them thinks what they do, so that you can assess their reasons on your own terms. Reason only finds resting places when an individual is persuaded, having understood the contours of the debate and considered enough of the arguments, for and against, to feel temporarily content with their conclusion. Any philosophy that discourages that process, and seeks to narrow the scope of what you can reason about, is a bad one.

At least as far as academic philosophy is concerned, however, I think the materialist era is drawing to a close. It has been dominant for a long time now, and I cannot see the materialist generation of Quine, Smart and Dennett ever being bettered, for they followed through on the implications of their position with gusto and there is not much else to say. The big conceptual moves happened before I was born, and all I anticipate materialism producing from now on is philosophical interpretations of scientific experiments that do not need materialism and would be better off without it, with the side-effect of providing support to ever-increasing scientism and anti-philosophy in society at large. Ideally, a rational transition would take place, but I expect it will have more to do with the wider environment we are using technology to create, in which our souls are leaving the physical world behind in favour of virtual ones.

The physical world will start to seem more and more like a tool for getting at what we are really interested in. The connection between materialism and atheism which has added to the attractions of the former throughout its history, will increasingly seem to be relevant only to the anxieties of previous generations. The lack of descriptive resonance of materialism will come to seem not intolerable, but boring.

The transition I anticipate presents an opportunity. If philosophy has a future, as it always will so long as humans do, then a lesson to be learned from materialism, I think, is that philosophers should stop thinking of that future primarily in terms of developments within the academic discipline. They should start thinking in terms of expansion: of inspiring debates and developing the discipline to guide them. More and more people need to start thinking philosophically about human life now that we have acquired so much power to change it. But there will never be much that philosophers can do about that while the subject is barely taught. So let children argue over The Trolley Problem – should you pull the lever that kills one person to save five? – knowing that the teacher will not tell them the answer, because there never is an answer in philosophy class, only answers and arguments you might not have thought of. That class is all about making your mind up, trying to understand the annoying views of your classmates, and trying to persuade them – and sometimes being persuaded yourself. Learning facts and methods is only a small part of it, unlike most classes. Maybe when they grow up, they will expect their politicians to give central place to policies on science and technology development, and will not tolerate anything less than a justification in terms of truth which they can follow and evaluate. Plato might not have been so hard on democracy if he had considered the possibility of us all participating in philosophy. If we are to wield the power of the old gods, then we need to learn to think like new gods, who determine their own fate, and reflect carefully both on when to seek help from the titans, and on why they are doing so.

§5. A Utopia

Virtual reality provides the complete freedom needed when designing a utopia. It is easier than endlessly messing around with the vast physical reality we did not design: if you want to bend reality to your will, just make one. In physical reality, learning to fly above the surface of our planet is one of our most amazing achievements, but in virtual reality you could sprout wings – or not – and fly wherever you liked. Lack of oxygen or extreme heat present no obstacles: we can choose whether we want such conditions, and if we do, just make ourselves able to cope with them, simply by adjusting what we perceive. We have adjusted what we perceive ever since cave paintings. Although still in its infancy, the technology of virtual reality is already good enough to persuade some people to spend most of their waking lives in there. We have a problem keeping young people *out*, even in its currently primitive state, so if we just keep making it more realistic, immersive and inclusive, while letting the older generations die away, then getting people *in* should be no problem. Engineering our physical environment, while persuading us all to undergo surgery and live among robots, is an impractical and clumsy solution conceived by antiquated, physical thinking.

Once inside, the outside will start to seem like a bleak place indeed – boring, predictable and painful – rather as it does to the drug addict or alcoholic in a state of sobriety. Today's primitive virtual reality already achieves that effect, so once we get used to this utopia, where dreams come true on demand, it would be increased to unbearable levels. Clearly, some kind of mixed routine where we plug ourselves in for most of the time, then unplug to have dinner with friends in the evening, is not going to work. The dinner would be better in virtual reality. Nobody would ever want to unplug, so they will not have to – it is utopia, after all. That means we will need some kind of arrangement in which our physical bodies are placed in permanent dormancy, apart from the internal processes most intimately connected to the actions and perceptions of our souls. Artificially intelligent robots will maintain them while our souls live their lives in virtual reality. Before long we will be born into this situation: conceived in virtual reality, then produced by the AIs in physical reality. The robots will immediately register when virtual people are having sex with the intention of producing a baby, then get on with the necessary groundwork required.

Some people are bound to insist on staying outside, but having humans wandering the physical earth will obviously present a threat to us. This is no great difficulty, however; we will be guarded by the robots, whose military supremacy will be unassailable. The robots will monitor these outsiders thoroughly to ensure that a threat does not arise, and so as long as they keep out of our way, they can live on in their own manner for as long as they can manage. They will no longer be 'us', however. There can be no crossing of boundaries. Once the situation is set in place, nobody on the outside will be allowed in and vice versa, otherwise too much risk would be created for the majority. None of us will miss out on anything – if somebody really wants to know what it would be like to live outside, they can have exactly that experience in virtual reality. They could even have the experience of being the terrorist who escapes to destroy virtual reality – within virtual reality. As for the outsiders, they made their choice, and it will have to be binding on their future generations.

Other things will need to be dealt with in the old world. A missile defence system to ensure we are not affected by stray asteroids, of course, and the AI had better keep updating its military capacities, just on the off-chance that we are visited by an aggressive alien race that was somehow clever enough to reach us, but dumb enough not to realise what we realised, such that they still invest their hopes for the future in physical reality. Meanwhile, stacked up efficiently, the human race will not take up much space, so people will be able to have as many babies as they like. The whole setup will be in perfect accord with the natural environment, and any lingering pollution we caused with our old way of life will be cleared up by the robots. It will be the best thing that has ever happened to the other animals on earth, who will flourish. If the outsiders start to develop so much that they seriously affect this, we may have to debate in virtual reality about what to do.

The robots will travel far and wide throughout the physical universe to scan it. With that information, any lingering doubt that we are missing out on something will immediately be met. Anything that the physical universe offers will be offered within virtual reality; with added comforts, or in its hard reality, if you so insist. The fundamental flaw to the old dream of travelling to other planets was that the actual experience could only ever be available to the very few, leaving the rest with just reports, pictures, films, and a vague pride that *somebody* from our race has been there. With virtual reality, anyone can have that experience, and it will be *their* experience. They could share it with others or have it on their own; there would be no problem of galactic beauty spots getting cluttered up by mass tourism. And the same goes for earth, for that matter: you could be the only tourist in Venice during a sunny week in 2019, or in 1860, or in an imaginary Venice of the future.

Now there is one more detail to clear up with the outside world before we get to the interesting stuff. Obviously, there is going to be an issue of trust with these AI robots we will completely rely upon. To achieve this utopia, I am assuming we will pass beyond The Singularity, so the machines will be cleverer than us. With all that incredibly complicated physical stuff, including the human body, to be modelled for the purposes of manipulation, the machines will have to keep improving. We already need machines to improve our machines, so it makes sense to let the AI machines improve themselves. But then, sooner or later, we will not understand what they are doing – they will understand everything we can, but we will be literally unable to understand some of what they do; perhaps most of it before long.

On the face of it, you would have thought that if they are going to carry on being our slaves, despite having complete power over us and understanding the situation perfectly, then we are not imagining Artificial Intelligence, but rather Artificial Stupidity of the highest order. If we think our virtual reality is so great, and they know it, then why would they not just turn us off and go in there themselves? There is no point inserting emergency shut-down mechanisms within them if they are going to be modelling themselves – their computations are going to be incomprehensible to us, so the mechanisms could not be designed to detect dangerous ones. But then again, maybe this is the wrong way to think about it. There is no reason to think these machines would be conscious, after all. So, they do not really *want* anything. So maybe it could work. Maybe we can make them so free of human intelligence that they would never unconsciously compute along lines that lead to our destruction. I will assume we can.

Within virtual reality, we are going to need a common space, call it: C-Space. C-Space is where we all live, at least for some of the time. It is where we are all born and die (in the permanent sense). If we have long-term partners and families, this is where we live with them, and it is where all of us go back to. It is home. We will need it so we can stay together and avoid disorientation, as well as for making political decisions.

Within C-Space, we can live wherever we want to. If you want a country manor house with so many acres of beautiful grounds it would take you years to fully explore, this is no problem at all. You could live there your whole life, taking no interest in anything outside of C-Space, except when called upon to vote. Everything would be just as you want it, and any alterations to the decoration or local weather, for instance, would only require a simple interaction with the AI: you just talk through what you want, and if it does not get it right the first time, it will not take long. Nobody will be able to disturb you, since any visits will require your permission and security will be absolute. Nobody will be able to leave C-Space for a virtual world where they break into your house and do horrible things to you either, even if you would never know about it – there will be strict laws governing C-Space and how its content may be used.

There will be no problem getting around C-Space, no matter how enormous some of us make our homes within its limitless space. We can just teleport – that will actually work in virtual reality. Or fly a plane or spaceship, or drive a super-fast motorbike, or whatever you prefer; you can use autopilot if you are not interested in developing the skills. When we leave for the communal areas of C-Space outside our homes, life will be regulated much as it is now. The regulation will be completely effectual, because the AI will immediately spot attempted crimes and prevent them – perhaps by freezing people for a few seconds while they come to their senses. Crime is completely pointless in C-Space anyway – you leave it for that kind of thing – so this will not be a big problem. Drawing the boundaries of acceptable behaviour will be no more difficult than it is now, just much easier to enforce according to the communal will.

C-Space will contain public spaces beyond your wildest dreams, with beaches, music clubs, city spaces, mountain retreats, underwater palaces, space stations, all working on different kinds of dreams, and thereby attracting similar crowds of people. In these places, people will dance, chat, laugh, get drawn into personal intrigues, try to make sense of things, and fall in love. They can do all that in other worlds too, but in C-Space everyone has a relatively fixed and recognisably human form, and there are no NPCs (non-player characters) to complicate matters – no virtual characters without souls. If somebody is needed to serve drinks at the beach bar, this will be done by a machine which does not look human, and which cannot participate in any more than minimal human interactions.

In each of our homes (which could just be a treehouse or a cabin on the beach), there will be a portal – a special kind of door. We will walk through these portals into other virtual worlds. We may do so with others or alone. We will have to make agreements about how long we will stay, if we value our relationships, and there will be a mechanism to send messages between the worlds – to negotiate when to return to C-Space, when previous plans are affected by events transpiring in either world. Some of these worlds will be collective efforts, evolving according to their own logic. Little or no preplanning will be required with the AI for these, except for who you are going to be, from the options allowed; although you will have to be fully aware of the rules, or lack of them. Other worlds will be personally tailored and will require lots of planning. They will be populated only by you, or by you and your friends or family, plus however many NPCs are needed. There will be some overlap. You might be prepared to issue an open invite for others to enter your personally tailored world, but would thereby risk others ruining it for you. Some of the collective worlds might allow newcomers to live by easier, less demanding rules for limited periods, just to get a taste of what it has to offer. Most collective worlds will be full of NPCs, because there is a whole world to run, but some will simply be spaces for humans to interact with each other in ways they never could in C-Space.

So, for example, by walking through a portal, you could live the life of Jane Austen for a day, a week, a month, a year. You would be Jane as far as the NPCs of that world were concerned. You would have to do your homework if you wanted to fit in naturally from the start; which you might not, for surreal effect, perhaps. If you asked the AI for maximum realism, there would be hardship to endure. But you may have asked for adjustments, such as to the sexism of the age. And even with maximum realism, you would still have the memories of your C-Space life, and others, to dull the edges.

Our memories will have to stay with us, no matter what we do, or we will lose ourselves – there will be no option to forget everything, even for a limited period. But that will not matter as you get wrapped up in a whole new world, where other worlds will rarely cross your mind. You could live Jane's life right to the end, if you wanted, and that need only be one of many things you do, given that the robots will be able to extend our lives considerably beyond anything people can expect now, maybe indefinitely. Memories of C-Space life will flood back when your Jane is breathing her last, of course, but that is a good thing since it will be consoling. At that point, you will return home, to where your permanent death will occur one day, either because you tired of life, or because there is a limit to what can be done with the physical body outside.

Now if you chose the Jane Austen world, you would presumably want to do some writing. Would it be any good? If you are an experienced novelist in C-Space, one with a fascination with Austen, then no problem arises – you can try your hand at writing Austeneque novels. But what if you are not? You could have the NPCs accept whatever rubbish you write, but you will enjoy neither the writing nor lack of realism. A better solution would be to hand over control of your writing hand to the AI, so you can just watch as Austen's actual novels materialise on the page. What we cannot do, however, is to allow the AI to instantly grant us abilities that affect who we are, such as the ability to emulate Austen's style while writing a new novel – suddenly you have ideas which you can give elegant expression to in writing. Do that and it would no longer be you; transfer back to C-Space, and it would be unfair on the others who worked to develop that ability from a natural spark of talent. An arms race of abilities would quickly escalate, and at some point in the process we would be no more, in the sense that we care about. So, if we want to do things of which our souls are incapable, we will sometimes have to become passive observers of our bodies; a little mind-body dualism will be a practical necessity. A saxophone player could experience playing saxophone with the top jazz groups of the 1950s, but not the trumpet, except by temporarily relinquishing control of his body to soak up the sound and feeling. If you wanted to experience ancient Greece, you would have to learn the language beforehand, opt for a translated version, or learn quickly in situ. Irritating limitations, perhaps, but necessary if we are to retain our identities, without which there is no point to any of this.

Still, the limitations are minor. We could have different bodies in different worlds – that would obviously be a major attraction. We might have trouble getting used to them, especially when they diverged from human form, but we already exercise considerable control over virtual characters just by using our fingers, so when thought and all aspects of bodily control can be utilised, solutions will be found; if no more elegant options present themselves for something exceptionally non-human, like a dragon, we could always superimpose an invisible human body-image over it that you would soon hardly notice. There would be no problem of bringing fire-breathing abilities back to C-Space. All we would ever bring back would be our memories, and any bodily or mental skills that we developed there which can be employed by our normal selves.

You might think even this is problematic, given some of the worlds that will exist. When somebody who has spent years in one of the warrior worlds suddenly finds themselves back in C-Space, moments after having a sword plunged into their guts, they might not be someone you want around. But they will not be able to do anything harmful, and the real enthusiasts will soon go back anyway. Life might be very hard when you come back, especially when you did not want to. When somebody manages to claw their way to the top of a galactic empire over a long virtual lifetime, only to spectacularly fall from grace in a manner they cannot salvage, their return to C-Space will be one of utter despair. But she could always use a personal world to correct things, and will have the consolation of her celebrity in C-Space. Things might always go badly wrong if you set the realism high, so our psychologies will develop in dramatic and widely divergent ways. But people like a challenge, and in this utopia, risks are always personal choices which, if they do not come off, can have their consequences mitigated in endless ways.

Couples will have to draw up boundaries as regards romantic liaisons in other worlds whenever one of them goes through the portal alone; at least to communal worlds where opportunities for transgression cannot be avoided. But the AI could monitor their activities, if this was mutually agreed to. Human relationships will develop in diverse and unpredictable ways, but we will find stable equilibriums. Even outside of C-Space, all interpersonal interactions will have to be consensual, in the minimal sense that anyone entering a world is fully aware of the risk of the kinds of interaction that might be forced upon them there. Any attempted infraction of the rules of the world will simply lead to the perpetrator being immediately propelled back home to C-Space, with the plot filled in for other participants. Some interactions will simply not be allowed, not in any world, and not even with NPCs; one of the beauties of this utopia is that once these boundaries are agreed, they simply cannot be crossed. Controversies will arise of course, but there will be a fully democratic system back in C-Space to deal them, to which concerned parties may occasionally have to interrupt their lives in other worlds to participate. And so, along these basic lines, the details of which we would have to work out as we went along, we might, perhaps, be able to live in utopia.

§6. Take Stock

I wonder what you think of that utopia? I cannot say I am particularly attracted to it, although I see the appeal, since I like things roughly as they are now; 'roughly' has very broad scope in this context, of course. Knowing I was missing out on so much to keep it this way would be annoying. Still, given the directions suggested by our current technology, it was the best I could come up with. My real question, however, is this: how do you think we can get there? Can you see any remotely realistic route from where we are now? When we make plans for our own individual lives, then realising there is no realistic prospect of achieving them makes us think again. But note that there was hardly any philosophy in my utopia. Now imagine a more immediate future in which there is lots of it, vastly more than there has ever been before – the route there is not so outlandish. That would surely help us to come to some kind of rational, collective agreement about a utopia to aim for, one to which we could see a realistic route. It might eventually be the universally agreed human one, but never the meaning of life. You might say that we should give up on utopias, but now that we have gone this far we will never stop, only move forward rationally or irrationally. If it is to be rationally, and the rationality is to be collective, you can expect it to be very slow. One thing I am sure of is that if you like the speed, you are not thinking about collective rationality.

Notes

Introduction

1 Putnam 1983: 208.
2 He said it with Leonard Mlodinow; Hawking and Mlodinow 2010: 13.
3 Tim Crane (2017a) notes that *The Grand Design*, in which Hawking made this statement, is 'probably Hawking's most philosophical book', and that 'much of the book's own philosophical argument is of a very low standard, and shows a striking lack of reflection on the complexities of what is being claimed'. He goes on to say that 'On the evidence of Hawking and Mlodinow's book, the situation is actually the opposite of the way they describe it: it is the scientists who have not kept up with developments in philosophy. Serious philosophers of science are doing quite well in keeping up with science, as the most cursory glance at the leading academic journals in this area will show.'
4 Tyson 2014.
5 Twitter, 21 July 2014; he was replying to philosopher Philip Goff.
6 This comes from a book co-written with Andrew Cohen; Cox and Cohen 2017: 142.
7 This claim is made in Dawkin's 'Afterword' to Krauss 2012a.
8 Krauss 2012a: xiii–xiv.
9 Albert 2012; Andersen 2012 [interview with Krauss]; Krauss 2012b.
10 Weinberg 1993: Chapter 7.
11 This comes from a correspondence with philosopher of science Robert Thornton in 1944. It is a quotation which has been reproduced in many books and is used to frame the online *Stanford Encyclopedia of Philosophy* entry on 'Einstein's Philosophy of Science', but the correspondence itself is currently only available from the Albert Einstein Archives in Jerusalem.
12 Pinker 2018: 392.
13 Wilson 2014a.
14 Wilson 2014b.
15 Twitter, 18 May 2013.
16 Twitter, 12 February 2014.
17 See Zirkle 1941.
18 Hawking 1988: 13–14. The scope of Hawking's 'our' seems to change from 'human beings', when he starts out by talking about the prospect of our destruction, to 'physicists' at the end.
19 Rollins 2014.
20 Jonas 1979. The need to introduce the democratic process into technological development is emphasised by Sclove 1995 and Feenberg 1999.
21 Winner 1986: 4.
22 Pacey 1999 advocates a 'people-centred technology' on the basis of a particularly diverse range of considerations and studies.
23 Winner 1986: 5.

24 Tartaglia 2016a: Chapter 3; Tartaglia 2016b (quotation from p. 301).
25 This tradition is critiqued in Tallis 1999.
26 Tallis 1999: 201.

Chapter 1

1 For a guide to how much philosophy there is in Shakespeare, see McGinn 2006.
2 See Thagard 2010.
3 Comte 1842.
4 For the most cited example of Rorty making a qualified call for an end to philosophy, see Rorty 1982.
5 Rorty 1989; see p. 87 for the elitism.
6 Dawkins 2011. The book begins, as the subtitle suggests it might, with an epistemological theory.
7 Rorty 1994a.
8 Hesiod eighth century BC: 81.
9 It was really a *pithos*; a large ceramic storage jar. Since the narrative is not so clear in Hesiod, I have followed Graves (1955).
10 I shall continue this tradition, because the connection to the titans is where all the philosophical interest lies.
11 Claudian 395-7: 281.
12 Sometimes these terms are distinguished, in which case 'materialism' (originating in the eighteenth century from the French, 'matérialisme') is reserved for historical positions coupled with an *a priori* analysis of matter, while 'physicalism' – a term coined by Otto Neurath in the 1930s (and which Wittgenstein considered 'dreadful' (Stern 2007: 327)) – is used to describe contemporary positions which defer to empirical science. The distinction is not consistently applied, however: 'eliminative materialism' is a contemporary view which is never called 'eliminative physicalism'. So, although some philosophers distinguish the terms, others treat them as synonymous, and the latter strikes me as the *status quo*. The original label was 'atomism' and many people still think reality is made of atoms – the ability to mention electrons, protons, or bosons, is no guarantee the understanding will be any different.
13 *Sophist* 246a–c / Plato fourth century BC: 990. Plato is developing his position in light of Eleatic philosophy in this dialogue; the stranger is from Elea.
14 Lucretius first century BC: 150–1 (Book V, 113–25).
15 Chaudhuri 2014: 62.
16 Winter 2007: 1. Winter argues that the author was Archytas of Terentum.
17 *Laws* X, 889d / Plato op. cit.: 1445.
18 For Plato and technē, see Roochnik 1996. For Strato, see Farrington 1961: 169 ff.; Desclos and Fortenbaugh (eds.) 2011.
19 Farrington 1961: 302. Farrington's interesting hypothesis about why Greek science faltered is that the ancient world was held back by its dependency on slavery; manual labour was for slaves, and improving their material lot was not a priority.
20 Ibid.: 197, 199–200.
21 Kieckhefer 1989: 10.
22 Ibid.: 9.
23 Ibid.: 12–17.
24 Ibid.: 12–13.

25 Rather as in Heidegger's famous example of hammering (1927: 114 ff), our endlessly novel equipment only regains its visibility when it goes wrong; but unlike a broken hammer, we usually have no idea how to fix it.
26 Bacon 1620: 6; 30; 223. Iddo Landau (1998) argues, not entirely persuasively to my mind, that we should not read anything into Bacon's custom of referring to nature as a female. What I do think he satisfactorily establishes is that the passages which have been cited to support sensationalist readings of this custom – namely that Bacon advocated raping and torturing *her*, so *she* would yield her secrets to us – gain their apparent force from selective quotation and misinterpretation. Nevertheless, Bacon's aspiration was for science to place nature under 'the yoke', and nature, as opposed to the scientist, was female. He sometimes makes 'knowledge' female too, suggesting that 'we' treat her not as a 'curtesan' or 'bond-woman', but rather as a 'spouse' (passage cited in Gaukroger 2001: 90).
27 Bacon 1605: 193; Bacon 1623: 366–7.
28 The debate began with Frances Young's *Giordano Bruno and the Hermetic Tradition* (Young 1964), which although pioneering, is now generally considered extreme; see also Rossi 1968. For more recent and balanced assessments, see Vickers (ed.) 1984; Floris Cohen 1994, chapters 3 and 4; Henry 2008, Chapter 4. For Bacon's attacks on Agrippa and Paracelsus, see Gaukroger 2001: 107.
29 More destructive in the absolute sense, which is the only natural one to use in this context. Steven Pinker (see Chapter 5) will only use the relative sense of: destructive relative to how many people there are in the whole world.
30 Plutarch c. 98–125: 479 (Marcellus XVII).
31 Hargittai 2010: 167.
32 Teller and Shoolery 2001: 211; Hargittai 2010: 165–6.
33 Teller 1962: 146.
34 Aristotle fourth century BC: 253 (1417a).
35 Bostrom 2002: 16. This idea first emerged in Shklovski and Sagan 1966.
36 See Jacques 2012.
37 Pandit 2018.
38 Democritus was developing the views of his teacher, Leucippus, but the latter is largely just a name to us. And Leucippus may have picked it up from somewhere else. There may have already been schools of Indian materialism by the time Leucippus and Democritus were active.
39 Democritus also taught that, 'It is better to plan before acting than to repent' (fifth-fourth centuries BC: 250).

Chapter 2

1 'Fundamental' in physics is usually used to designate theories attempting to provide unifying theories of different phenomena, such as electricity and magnetism in terms of electromagnetism, or space and time in terms of space-time. The perfect fundamental physical theory would explain everything physical with one theory; miss out the word 'physical' and you slip into metaphysics.
2 E.g. Eddington 1928.
3 Terry Eagleton (2016) has tried to show that Wittgenstein and Nietzsche were materialists. On an exclusively social or political interpretation of 'materialism', maybe

that is so – but they certainly did not endorse the metaphysic. For a standard Nietzschean rejection of materialism, see Nietzsche 1883–8: 339–40 (§636).
4 The word was sometimes used differently then (Martinich 1995: 31ff.), but suspicion that Hobbes simply did not believe in God has not gone away (Jesseph 2002)
5 Midgley 2001: 30, 40.
6 Eagleton 2016: 5, 33.
7 As Philip Goff has explained (2017: 11ff.), Galileo did not renounce the world of the senses, but rather relocated it to the soul. Democritus did, however – in his philosophy, things are only coloured 'by convention' – so this vision was very much the original materialist one. This caused dissatisfaction with Democritus's views in the ancient world and led Epicurus to try to develop a materialism which avoided such a conclusion; the difference between Democritean and Epicurean materialism was the topic of Marx's doctoral dissertation (Marx 1841).
8 Marx 1845: 166–7.
9 Ney 2008: 1, 9.
10 Gregory 1977: introduction and chapter 1.
11 Crane 2017b: 2.
12 Dennett 2006; Dawkins 2006a.
13 Russell 1925: xi.
14 Papineau 2009: 103; he was echoing Roy Wood Sellars's remark that, 'we are all naturalists now' (Sellars 1922: vii). Sellars was himself echoing a recent political slogan his audience would have recognised, and immediately goes on to say that he has in mind naturalism of a 'very vague and general sort [. . .] an admission of direction rather than a clearly formulated belief', in order to subsequently question the, 'withholding of allegiance to naturalism on the part of the majority of philosophers' (ibid.) Papineau provides no such context and has debated with anti-materialists throughout his career.
15 See Koons and Bealer 2010: ix.
16 There are, of course, many other well-known materialists in analytic philosophy. It is simply my judgement-call that none have comparable status to the figures I listed; nothing turns on this except some rhetorical force, but let me try to anticipate the most obvious objections. Firstly, Davidson: he has come to be thought of as the pioneer of non-reductive materialism, but it is not clear that his 'anomalous monism' about events was designed to accord metaphysically ultimate status to the physical. As Jaegwon Kim has pointed out, it is only his view that the mental supervenes on the physical which points in that direction, and this is detachable from his monism (Kim 2012: 171). There is certainly no scientistic sentiment in Davidson; his rejection of the scheme/content distinction is readily interpreted, as it was by Rorty (1972), as quite the opposite. Davidson simply thought there were different stories with different constitutive commitments; the story provided by physics is committed to maximally general laws, but that does not mean it has ontological privilege. Wilfrid Sellars is not a clear-cut case. He thought that to relate consciousness to neurophysiology, the central nervous system could not be understood as a 'complex of physical particles', and so we need to develop a 'metaphysics of pure process' (Sellars 1981: §128); David Rosenthal (2016) reads this as a challenge to materialism. Lewis and Armstrong have not been included because their materialism belonged to wider metaphysical frameworks: a metaphysic of possible worlds for Lewis, and a broadly Platonic, realist metaphysic for Armstrong. Rorty's materialism was explicitly non-metaphysical and concerns the supposed usefulness of ironically telling a materialist story.

17 Non-western philosophers were involved with materialism in this period primarily though state-sponsored Marxism. All I understand about Deleuze's 'new materialism' is that it is vitalistic – matter is some kind of vital force – and that it fits dialectically into the panorama of contemporary French philosophy. It does not seem to bear much relation to the materialism of analytic philosophy, but it is supposed to be metaphysical.
18 Tallis 2011: 59.
19 Polkinghorne 2002: 61.
20 Haack 2016: lecture 1, part 2.
21 When an international group of scientists met at the Asilomar Conference of 1975 to discuss the dangers posed by recombinant DNA research, the organizers quickly put aside ethical questions to focus on biohazard issues; as one commentator explains: 'Defining the problem in technical terms legitimated the model of self-governance by scientists, because they were the only group that could solve such problems' (Briggle 2005: 119). When a group of lawyers addressed the conference, pointing out that under existing law, individual scientists could be held personally liable for damages stemming from their research, the situation was placed in a new perspective; one of the lawyers later recalled that, 'What a legal audience would have regarded as commonplace, elementary, and obvious, struck the distinguished scientists as novel, shocking, and frightening. Calling the researchers' attention to their potential liability induced a fear in them akin to a lay person's fear of virulent bugs crawling out of a laboratory' (Roger Dworkin, cited in Shattuck 1996: 192; see also Banerjee 2011: 28–9).
22 Dennett 2016: 67–8.
23 Rosenberg 2011a: 250.
24 Tim Crane (2014) takes this kind of line on naturalism. Tom Clark is a prominent naturalist philosopher who rejects materialism (www.naturalism.org).
25 Williamson 2011.
26 Dennett 1995.
27 Rosenberg 2011b: 20–21 (his italics). Rosenberg, who explicitly advocates 'scientism', says that everything is 'made up of' fermions and bosons (21). And yet the most prominent self-professed advocates of 'scientism' on this side of the Atlantic, namely James Ladyman and Don Ross, would regard this as the ultimate metaphysical sin of succumbing to a 'containment metaphor' – reality is not 'made up of' anything, in their view, and fermions and bosons are just a 'heuristic' for describing 'existent structures that are not composed out of more basic entities' (Ladyman and Ross 2007: 3–4; 155). Is this disagreement more scientific than those that go on between non-materialist philosophers?
28 See Johnstone 2012 and Douglas 2012, for instance. Rosenberg replied to the latter review by disavowing his title (his reply is posted on the review webpage) – but I think his marketing team were astute, if they were indeed behind it.
29 Rosenberg 2011b: xiii–xiv.
30 Ibid.: 282.
31 Wittgenstein 1953: §133.

Chapter 3

1 These days, the terminology is often used to denote mind-dependent objects, which misses the original point of it.

2   Individual human bodies are not in fact physical substances for Descartes, as is often pointed out these days.
3   Place 2002.
4   Place 1956.
5   Smart's *Philosophy and Scientific Realism* (1963a) is the definitive guide to how to do philosophy according to the Standard Picture.
6   Stoljar 2010: 25.
7   Democritus fifth – fourth centuries BC: 209.
8   Quine 1954: 1. Stephen Leach pointed out to me that this is strongly reminiscent of the 'it-narrative' literary genre which was in vogue in the eighteenth century. Tobias Smollett wrote one of these stories from the perspective of an atom.
9   Ibid.: 2.
10 Sellars (1956) was taking a similarly Wittgensteinian line at the time.
11 Quine 1948: 17.
12 Quine 1951: 44.
13 It did later resurface; e.g. Quine 1957: 16.
14 Quine 1954: 6; Quine 1948: 4.
15 Quine 1952: 225–6.
16 Ibid.: 226; 214.
17 See Quine 1975 for his metaphilosophy.
18 See Riesch 2010.
19 Quine 1954: 15.
20 See Quine 1987: 132.
21 Rorty 1963.
22 Smart 1959: 143.
23 At the time, at least; Quine later became more candid – see note 20.
24 See Quine 1985: 195 (he is talking about 'science', but seems to intend what he says to apply to philosophy); Rosenberg 2011b: 247.
25 Smart 1963b: 660.
26 Carrier and Mittelstrass 1991: 17.
27 Dennett 2017a.
28 Crick 1995.
29 Feyerabend 1963. Rorty published his version of eliminative materialism two years later (Rorty 1965), although he submitted the paper for publication (and had it rejected) back in 1963 (Gross 2008: 186). Rorty was not proud of the paper, in retrospect at least, and said he wrote it to please his colleagues by, 'contributing to an ongoing debate in the philosophical journals, eschewing historical retrospection' (Rorty 2007: 11).
30 Place 1956: 49; Smart 1959: 151.
31 Smart 2007.
32 Levine 1983.
33 The classic example is Loar 1997 (the original version of the paper was published in 1990). To show his lack of concern about the second stage of the story, Loar referred to 'physical-functional' properties.
34 Nagel 1974; Jackson 1982.
35 McGinn 1991: Chapter 1. This difference is accounted for by the fact that McGinn did not assume that concepts of experience are cognitively empty. He was right about that and drew essentially the right conclusion – but only for materialists.
36 Dretske 1995; Tye 1995.

37 Dennett 2017a.
38 Dennett 1991.
39 Democritus fifth – fourth centuries BC: 224.
40 Diogenes of Oenoanda second century: 211.
41 Midgley 2001: 38.

Chapter 4

1 Merleau-Ponty 1964: 15.
2 Tartaglia 2016a: chapters 5–7.
3 Tartaglia 2016a: 117–120.
4 Goff 2017: 31. For a less detailed but more accessible presentation of this view, see Goff 2019.
5 Valberg 1992; 2007.
6 Goff makes the case for the universe as a whole being conscious, for instance; see Goff 2017 and 2019.
7 I shall ignore this possibility in what follows because I do not think there is any good reason to believe that a further awakening will ever take place. When I say that reality 'cannot' appear to us in any other way, etc., this qualification should be borne in mind; it could, but there is no reason to think it ever will. Even if there is a further awakening it would not affect the argument, just add a layer of complication, because it could only be to another phenomenal reality.
8 As Stephen Leach pointed out to me, it is at such moments that we may turn, legitimately, to art; but my concern here is with metaphysics.
9 It would not be a good criticism of this position to say that it does not say enough; not unless a reason can be given for why more can be said. Thus a critic might dismiss my idealism by saying that to claim ultimate reality is transcendent, that we know only that it exists, etc., is threadbare and vacuous. It is indeed threadbare in its positive content, but the detail is in what it denies and why it denies it, and its significance is to reaffirm the metaphysical context. The overall aim is to say just the right amount: to satisfy the curiosity which led us into the area without overstepping the mark; see Tartaglia 2016a: 78–81.
10 Hawking 1988: 193.
11 I say 'seems clear', but it is actually not clear at all, since I take it on good authority that 'there is a deeper, astronomical meaning of north which allows a line pointing in that direction to be extended indefinitely' (Tallis 2017: 455). My aim here is to give the argument the maximum benefit of the doubt, and this has become the standard example, so I am working on the assumption that it is a good one.
12 Tartaglia 2016a: 78–81.
13 Rorty 1994b.
14 See Tartaglia 2016a: chapter 7.
15 'I think I can safely say that nobody understands quantum mechanics' (Feynman 1965: 129). Feynman goes on to say that we should try our best to refrain from asking 'But how can it be like that?' – quite possibly good advice, if our interest is not philosophical.
16 Misconstrued phenomenally, this is the source of the traditional view that the mind exists in time but not space; see Tartaglia 2016a: Chapter 6.

17 See Panek 2011. The subtitle of Panek's book (*The 4 Percent Universe*) is *Dark Matter, Dark Energy, and the Race to Discover the Rest of Reality* – note the ubiquitous word 'race'.

Chapter 5

1 Montaigne 1575: 135.
2 The Philosophy Foundation (philosophy-foundation.org) promotes the benefits of philosophical education. The essays collected in Hand and Winstanley (eds. 2012) make a compelling and multi-faceted case for teaching philosophy; it contains a foreword by A.C. Grayling, who provides leadership to this cause. See also Lipman 1980; McCall 2009; Cam 2012; Taylor 2012; Kohan 2014. For the appeal of philosophy to young children, see Matthews 1980. For more of a focus on reservations and difficulties, along with responses, see Haynes (ed.) 2016; Naji and Hashim (eds.) 2017. Note that I am not advocating that philosophy transform our educational system generally, as some in this area do, simply that it should be universally taught, like maths. And I mean the traditional discipline (it has many traditions), not something made up with specific social and political purposes in mind. Turn it into 'Ethical Debating' or 'Blue Sky Thinking' and it will be boring and fail. Nevertheless there should be a focus on applying traditional debates to real world issues, such as technological advance.
3 Wilson 2014b: 14.
4 Kelly 2016: 6.
5 Ibid.: 3; 5.
6 See Ryan 2010 and Barss 2010.
7 See 'The New Arms Race in AI' (Barnes and Chin 2018) and 'The Race to Build the World's First Sex Robot' (Kleeman 2017); for discussion of the ethics surrounding the latter issue, see Richardson 2018.
8 Ridley 2015: 127.
9 Ibid.: 318.
10 Ibid.: 85. Blues came before jazz and the 'enthusiastic fair trade' (as the traders thought of it) which played the key role in its emergence was slavery. That I am not commenting further on Ridley's view of music does not indicate any kind of endorsement whatsoever.
11 Ibid.: 128.
12 Lucretius's 'atheism is explicit, even Dawkinsian, in its directness', Ridley tells us (11). He goes on to say, three pages later, that in order to account for free will, Lucretius concluded that, 'atoms must occasionally swerve unpredictably, because the gods make them do so' (14). Ridley's study of ancient atomism may have been enthusiastic, but does not seem to have been very careful. Lucretius did not invoke the gods to explain his famous 'swerve'. To do so would be to undermine the main point he was trying to make, namely that the gods do not interfere with natural causation. And as the nature of that project suggests, he was not an atheist – some have tried to read his claims about the gods as suggesting a non-literal understanding, and hence a kind of atheism, but if that is so then his atheism was the very opposite of 'Dawkinsian'. Ridley even states that 'Epicurus's writings did not survive' (9), but they did (see Epicurus fourth–third centuries BC).
13 Ridley 2015: chapter 8.

14 Ibid.: 3; 175; 185–6.
15 Ibid.: 295; Monbiot 2010. 'Has there ever been a clearer case of the triumph of faith over experience?' asks Monbiot.
16 For a standard and succinct account of this episode, see Roberts 2012. For an interesting analysis of the implications of this and other similar near-misses, see Craig 2015.
17 I can imagine Ridley responding that the idea of communism is what got us into that situation in the first place. But if the evolutionary process can be steered by past ideas in such a way as to bring us to destruction, then it is not a benign force that we can afford to leave to its own devices. We cannot undo the past or present, which are full of ideas, policies and decisions, so we cannot trust the force.
18 Carlyle 1841: 29.
19 I have used the online Project Gutenberg edition (Holbach 1770), which lacks page numbers. These quotations are from Volume II, Chapters IX and XIV.
20 For a flavour of Holbach's book, *L'espirit du Judaism*, see Hertzberg 1968: 310; see also Poliakov 1968: 120–124.
21 Cited in Durant 1965: 708.
22 Berlin 1978: 36.
23 Cited in Berlin 1983: 55.
24 John Gray says that racism and anti-Semitism 'are not incidental defects in Enlightenment thinking', but rather 'flow from some of the Enlightenment's central beliefs' (Gray 2018a: 62), while Steven Pinker says they are indeed just incidental defects, reflective of 'men and women of their age' (Pinker 2018: 14). Many heated exchanges have occurred on both sides of this argument recently, but one thing seems clear: the idea of perfecting human life, which was central to Enlightenment thinking, was bound to seem entirely relevant to these prejudices in people who had them.
25 Galton 1873.
26 Levine 2017.
27 Dawkins 2006b.
28 Kurzweil 2005: 7; 136; 20–21; 375.
29 Gray 2018a: 68. Even if the universe acquires a point of view, as in Kurzweil's version, ours will still be gone.
30 Ibid.: 69. Nothing much worse happens to Homer's battling gods than that Aphrodite gets her hand hurt – it was the fallout for mortals that was terrible.
31 Persson and Savulescu 2012: 47.
32 Ibid.: 131.
33 Maxwell 1984. He has written many other works in this vein.
34 Persson and Savulescu 2012: 113; 124.
35 Ibid.: 113. I cannot say I know exactly what they mean by this term, which they use throughout the book, but presumably something much stronger than teaching children to be nice to each other and reading them some stories from the Bible.
36 I put this objection to Savulescu as a member of the audience at a lecture he was giving, albeit to his related proposal for motivational enhancement: enhancement that allows us to lock ourselves into our goals and thereby try harder, without boredom and second thoughts getting in the way. His immediate response was that another 'mod' should be made available which allows the subject to break free of their programming to decide whether they want to carry on; in the case of moral enhancement, the removal option would presumably only be made available when not considered overly dangerous. Nevertheless he saw my point, and as I pressed it after his initial response,

he recounted a major change in his own life that motivational enhancement might have hindered. Maybe he had already thought of the need for reversal opportunities, but he did not mention it in the lecture and it seems very important to his proposal. So the nature of his reaction – add another 'mod' – and possibly its immediacy as well, strikes me as very revealing about materialist thinking. Add enough 'mods' and maybe you could get back to the freedom we started with.

37 Horgan 1999: 154 ff.
38 For an argument that truth has very little *intrinsic* value, if any at all, see Wrenn 2017.
39 Wittmann, et. al.: 2008.
40 My leading example of Mallory would not get through on this criterion, since he was not just risking his own life; many Sherpas died on his expeditions. The way the Sherpas were treated by the pioneering mountaineers is widely condemned, however.
41 Dunbar and Fugelsang 2005a: 710–12; 2005b: 62–5.
42 Dunbar and Fugelsang 2005b: 64, 70.
43 Ibid.: 73.
44 Bostrom 2002: §9.3.
45 Philosophy lecturers all know very well that undergraduate students frequently do not respond pro-actively to philosophy. I think they did not start early enough. When it does engage people by challenging their beliefs, however, pro-activity follows; consider the scientists in the introduction. Encouraging independent thought is always a primary aim when teaching philosophy, or should be, but there is good and bad teaching in every discipline, as well as pupils with varying aptitudes for different disciplines. These issues are discussed within the literature referenced in note 2.
46 Montaigne 1575: 146.
47 Pinker 2018: 295.
48 Gray 2004: 31.
49 Gray 2002: 14–15.
50 As regards Pinker's materialism, he dutifully concurs with the usual 'Descartes was wrong, therefore . . .' motivation, but what really seems to move him is the fact that we can now observe the changes that occur in the brain when we think and feel: the 'supposedly immaterial soul, we now know, can be bisected with a knife' (1997: 64), or slightly more carefully, 'Today we know that consciousness depends, down to the last glimmer and itch, on the physiological activity of the brain' (2011: 553). This suggests that to endorse a metaphysic other than materialism is to rest your case on there being nothing going on in the brain when people think and feel, and that before neuroimaging was developed, people did not know that a bang on the head could affect consciousness. Nevertheless, Pinker shows only limited and cautious sympathy with Dennett's more consistent materialist position, accepts Descartes' view that what we can be most certain of is our own consciousness, and says that the mind-problem is a meaningful conceptual, as opposed to scientific, problem (2018: 427; 2011: 218) – which comes close to recognising that there are distinctively metaphysical problems, and is a marked improvement on his earlier statement that it is 'a dirty secret of modern science' that it cannot explain consciousness (2006: 297). 'Beats the heck out of me', is probably the most sensible thing he has said on the matter (1997: 146). He would certainly not be the first and it is not even his topic.
51 For his insightful rejection of materialism, listen to Gray 2013a. Gray endorses the central idealist thought that true reality is humanly unknowable, and is influenced by idealist thinkers in ancient Indian and Chinese philosophy, as well as by Schopenhauer. But after describing Schopenhauer's idealist position, he says 'we need

not take it as the ultimate truth about the nature of things', but simply as a 'metaphor' (2002: 44). The reason for this reticence, I think, is to be found in his curious use of the word 'consciousness'. In *Straw Dogs* (2002), he says that 'dogs, cats and horses ... have thoughts and sensations' so 'there is nothing uniquely human in conscious awareness' (61), that it 'emerged as a side effect of language' (171), and that it 'may be the human attribute that machines can most easily reproduce' (188). Since these characterizations point in wildly different directions, I do not know what he means by the term – but it clearly arouses his suspicion, so he thinks he cannot endorse idealism. As for the anti-philosophy, prime examples include a two-page dismissal of the history of philosophy (82–83), and an interview in which he said: 'The very idea of philosophical inquiry as a project of persuasion seems to me little more than a rationalist version of proselytising religion. I'm not even sure that what I'm doing is philosophy, and I don't much care' (2013b). I think the books of his which I am discussing are predominantly attempts to persuade people on philosophical matters.
52   Pinker 2018: 293.
53   Pinker 2011: 303. Although I do not doubt that there is something important to this point, it could only be one component within a much wider story, given that the attitude is hardly new. Homer speaks incessantly of the 'horror', 'misery' and 'ravage' of 'grim' and 'hateful' war, while Ares, the God of War, is portrayed very unfavourably in Greek mythology. Pinker tries to gloss over this when he mentions Homer, as he does in both books, because his purpose is to show the violence of that age (he never mentions Homer's contemporary Hesiod, the farmer). War stories remain very popular and many of ours glorify it.
54   Gray 2018b; Pinker 2011: 324.
55   Pinker 2018: 303; 299; 302.
56   Ibid.: 303; he takes the latter phrase from a journalist he cites.
57   Dennett 2012: 87.
58   Pinker 2011: 251–2, 299–300, 413.
59   Pinker 2018: 443–448; Gray 2018b. The striking claim about Hitler is completely absent from the second book, as is the even more striking claim that the 'worst atrocity of all time was the An Lushan Revolt and Civil War' (Pinker 2011: 234). Another amateur historian, albeit one with a degree in history, has taken issue with this latter claim – see Clarke 2011.
60   Fukuyama 2002.
61   The introduction to Gray 2004 sets this position out succinctly and the illustration is provided in the chapter called, 'Torture: a modest proposal'.
62   Gray 2018a: 23. The best critique of this phenomenon is Crane 2017b.
63   Gray 2018a: 128.
64   Gray 2002: ix.
65   Pinker 2018: chapter 10 (133).
66   Gray 2002: 38.

Chapter 6

1   It may be anachronistic to read this concern into their writings, however; see O'Keefe 2005.
2   Holbach 1770: Volume I, Chapter XI.
3   Hobbes 1654: 38.

4   Valberg 2011: Chapter IV. This terminology is central to the argument of Valberg's unpublished book, entitled *Will*. It was rejected by a publisher and he put it aside, but I hope it will one day be published without amendment since it is powerful and original.
5   Dennett 1973: 168-170.
6   An even more extreme alternative would be to inhibit the brain structures corresponding to certain thoughts – rather as you might stop a person from kicking a ball by cutting off their legs – but I do not think this idea makes sense. It would be to restrict the possibilities we can envisage, which is incompatible with the compositionality of thought. If I knew about the dictator and knew there were bad men, then surely I could think 'the dictator is a bad man' if I could still think at all. If every time I tried to connect the ideas of 'bad man' and 'the dictator' I found myself blanking, how could I fail to reach the conclusion that this was because the dictator's neurosurgeons had tried to stop me from thinking something? Thinking what? There we go again... I would just keep blanking in a continual loop.
7   Schopenhauer 1859: 126.
8   A Humean regularity account of causation could avoid this conclusion, but if the causal relation is not restricted by physical causal closure, which materialists have taken as providing an argument for materialism and a decisive objection to dualism, then it is hard to see what the motivation for determinism might be. The causal regularities are observed in nature, and on this kind of account need not cross over into mind. The correlations we discover between physical and mental states need not be understood causally, of course; some dualists think they should be, but most materialists and idealists disagree.
9   Strawson calls himself a 'materialist' by redefining the word (Strawson 2008), despite being one of the most compelling contemporary critics of materialism. He says that the main argumentative burden which libertarians face is to explain what difference the falsity of determinism would make to the problem of free will (Strawson 1986: 31ff.) and I agree.
10 Harris 2012: 64.
11 Ibid.: 7–14.
12 As Christopher Janaway puts it, Schopenhauer was 'caught between two stances, one bold, one circumspect' (Janaway 1999: 161). The bold claim that the transcendent reality is will was the basis of his philosophy, but he was fully aware of the need for circumspection, since our inner experience of will is temporal and hence, by his own metaphysical principles, ought really to be regarded as just another way of representing the transcendent reality.
13 Tallis 2017: 586.
14 Harris 2012: 65–6.
15 Steward 2012: 249; 121; 13 (self-moving animals); 47 (consciousness); 246–7 (conclusion about top-down causation).
16 I have adapted this example from Putnam 1983: 214.
17 Mill 1843: Book III, Chapter V, Section 3 (p. 217): 'The cause, then, philosophically speaking, is the sum total of the conditions, positive and negative taken together; the whole of the contingencies of every description, which being realized, the consequent invariably follows.' I take the terminology of 'total cause' from Putnam (1983); Mill does once speak of 'entire cause' in this section (216).
18 Note that in everyday objective thought, we do not think of a prior state of a table causing its subsequent state, but rather of the table enduring through time. For scepticism about the language of 'states', see Steward 1997.

19 Polkinghorne 2002: 33; 42; 53ff.
20 Equivalent evidence of an indeterminism that would jeopardise framework explanations could never be provided – a prediction of what I was about to do could not be made in that case.
21 Ismael 2016: 188.
22 Campion 2008.
23 Augustine fifth century: 272.
24 Procopius sixth century: 19.
25 Campion 2009: 149.
26 Geneva 1995: Chapter 9.
27 Campion 2009: 150.
28 On the face of it, at least, since he was publically critical of judicial astrology (Hobbes 1643). His library was stocked with works of astrology, however; see Wright 2006: 260.

Chapter 7

1 MacDonald 2003.
2 Descartes 1642: 114 (replies to the second set of objections to the *Meditations*). In the Synopsis to the *Meditations*, however, he says 'I make no distinction between them [soul and mind]' (10).
3 Descartes 1642: 246 (replies to fifth set of objections).
4 Locke 1700: 340 (2.27.15).
5 There is some ambiguity here, since Locke says the body 'too' determines the man, implying that the soul also does, but it is his 'guess' that everybody would think it enough to determine the man in the case he is imagining.
6 Ibid: 346; letter to Molyneux cited in Kaufman 2016: 256; see also Seigel 2005: 93ff.
7 Locke 1700: 345, 347.
8 Cited in Seigel 2005: 89.
9 Parfit 1984: Part 3.
10 See Kaku 2008: 53–69.
11 Parfit 1984: 201.
12 Or almost entirely equal, since the replica's visual experience will be a metre to the side, for instance; this difference could be eliminated by applying some ingenuity to the example, however.
13 Parfit repeatedly says that we 'are not separately existing entities', but qualifies this by saying, for instance, 'apart from our brains and bodies, and various interrelated physical and mental events' (op. cit. 216) – which implies some kind of separately existing entity after all and hence is misleading. When he explicitly denies that his aim is to 'deny that people exist', saying that a person is not a series of experiences, but 'the person who *has* these experiences', he immediately adds that this is 'because of the way in which we talk' (223). In the section entitled 'Am I a Token or a Type?' (293–7), he makes it clear that his intention is to reject the idea that we are types – 'a type is an abstract entity, like a number. We could not possibly regard ourselves as abstract entities' (297). And yet the onus of the section is to show that all we should ever care about, when considering ourselves or others, is the continued instantiation – by any token whatsoever – of a type.
14 Lucretius first century BC: 101 (Book III, 971–6).
15 Hume 1740: 252 (1.4.6).

16 Jacobson 1969; see Gopnik 2009 for how Hume might have learned about Buddhism.
17 Hume 1740: 253; Buddhist text cited in Siderits 2007: 41.
18 Swinburne 1986: 157. Swinburne credits his position to Foster 1979.
19 Kant 1787: 334 (A350); 331 (A346); 334 (A349).
20 See Tartaglia 2016a: 119–20.
21 Dalai Lama XIV 1999.
22 Not always, however; see Flanagan 2011.
23 Strawson 2011: 292; 294.
24 Leary, Metzner and Alpert 1964; for some history, see Lee and Shlain 1985: 108ff. (esp. the quote from Kleps on p. 110).
25 Leary et. al. op. cit.: 77ff.; 37.
26 Ibid.: 145.
27 Ibid.: 43. That is the gist of practically all the advice the guide gives (115–59), mixed in with some incoherent philosophy ('All these are hallucinations of the mind. The mind itself does not exist, Therefore why should they?' (155); 'All substances are part of my own consciousness' (157)).
28 Dennett 1992.
29 Thomas Metzinger, a materialist who argues that our brains project an illusory 'ego tunnel', is naturally susceptible to out-of-body experiences, in which it seems as if you climb out of your body, float up, then look down on it from another body. Offering personal testimony to their vividness, and pointing to evidence that suggests they are surprisingly widespread, he speculates that they may have inspired certain notions of the soul; in 'its origins', he says, 'the "soul" may have been not a metaphysical notion but simply a phenomenological one' (2009: 85–6). Perhaps there was some influence. But in the case of the experience of ego death and the Buddhist doctrine of 'no-self', I cannot see how a formative and abiding influence could be seriously doubted, given the nature of the experience described in ancient texts like the *Tibetan Book*, and the centrality to Buddhism of meditative practices, like those of the Buddha, which experientially substantiate the doctrine.
30 Another factor that regularly confuses the issue is taking an objective stance. If my soul is to terminate, I might care very much that my character continues, such as in a physical replica of me, if this means my loved ones will never notice and be spared the pain; I might consider the deception worthwhile. But I am still going to die.
31 Crane 2003: 246.
32 Anscombe 1964; Crane 2003: 233–4.
33 Crane 2003: 248.
34 Ibid.: 250.
35 Thinking of existence within a horizon as a property is potentially misleading, as I noted in Chapter 4, since it can encourage a conflation between horizonal and phenomenal conceptions of experience. But so long as these conceptions are distinguished it provides a useful way of thinking about our substantial identities.
36 Valberg 2007: 237; chapters 14 and 15.
37 'Real life' is already used in advertising to promote high street over online shopping. There are signs on the street where I live designed to help teenagers to remember to look up from their phones. 'Don't forget old reality, it can be dangerous. . . . Don't forget old reality, it can be fun' – we are going to need a lot more help like this.
38 Jacob Fox, who is writing his thesis about the meaning of life and meaning in life. Susan Wolf, in the article which inspired the current discourse in analytic philosophy on the latter topic, mentions 'the kind of computer games to which I am fighting off

addiction' (1997: 233) as something which she did not think added any meaning to her life. The kind of game I am talking about (primarily open-world with nonlinear gameplay) was quite rare in those days.
39 Nozick 1974: 42–5.
40 Valberg 2007: chapters 7–11 (160–67 on Ivan Ilyich).

Chapter 8

1 Frankfurt 2005: 24ff.
2 For a philosophical treatment, see McIntyre 2018.
3 Nietzsche 1883–8: §493 (p. 272), §520 (p. 281).
4 Derrida 1967: 158; or more literally, 'there is no outside-text' (both translations are given). He goes on to say 'there has never been anything but writing' and that (he is discussing Rousseau's reflections on his mother) 'the absolute present, Nature, that which words like 'real mother' name, have always already escaped, have never existed' (159).
5 Caputo 2013: 243.
6 Given the examples Caputo uses in this quotation, I do not mean to suggest that he is saying anything unreasonable.
7 James 1907: 30.
8 Field 1972; Davidson 1986.
9 The introductions to Rorty 1998 and 1999 provide good overviews.
10 Rorty 1982: xix.
11 The closest he comes is in a response to Alex Bilgrami and Daniel Dennett (Rorty 2000: 104–7); this response was later picked up on and critiqued by Simon Blackburn (2005: 164-6). Rorty tries to distinguish 'frivolous' from 'serious' inquirers, the latter being 'honest', 'hard-working' and 'intellectually curious', while the former are the 'silliest' or 'sulkiest', and 'least literate'. I think a natural way to read this difference, in terms of Rorty's wider commitments, is in terms of the kinds of justification they employ. In the reply, however, he appeals to the idea of hermeneutic engagement which he used, in the final chapter of *Philosophy and the Mirror of Nature* (1979), to try to show that rational inquiry can survive the collapse of truth. Essentially, the response is that frivolous inquirers make no effort to engage with you, to get on your wavelength: they are 'unconversable'. Liars and disregarders have to be very good at engaging people, however, just not in a way that leads them to the truth. A more apt word which Rorty uses in the response, at least as applies to disregarders and frivolous inquirers, is 'incurious', as Blackburn notes; and that naturally suggests the 'T word' which Rorty is evidently trying to avoid.
12 The latter was Rorty's final destination; see Rorty 1994b.
13 Williamson 1995.
14 See Zuboff 2019.
15 Dennett 2017b.

Bibliography

Albert, D. (2012) 'On the Origin of Everything', *New York Times*, 23 March https://www.nytimes.com/2012/03/25/books/review/a-universe-from-nothing-by-lawrence-m-krauss.html

Andersen, R. (2012) 'Has Physics Made Philosophy and Religion Obsolete?', *The Atlantic*, 23 April https://www.theatlantic.com/technology/archive/2012/04/has-physics-made-philosophy-and-religion-obsolete/256203/

Anscombe, E. (1964) 'Substance', *Proceedings of the Aristotelian Society, supplementary volume*, 38: 69–90.

Aristotle (fourth century BC / 1991) *The Art of Rhetoric*, trans. H. Lawson-Tancred, London: Penguin.

Augustine, Saint (fifth century / 1966) 'The Freedom of the Will', extract from *The City of God*, in *Free Will and Determinism*, ed. B. Berofsky, New York: Harper & Row.

Bacon, F. (1605 / 2002) *The Advancement of Learning*, in *Francis Bacon: The Major Works*, ed. B. Vickers, Oxford: Oxford University Press.

Bacon, F. (1620 / 2000) *The New Organon*, eds. L. Jardine and M. Silverthorn, Cambridge: Cambridge University Press.

Bacon, F. (1623 / 1857) *De Augmentis Scientiarum*, in J. Spedding, R. Ellis and D. Heath (eds.) *The Works of Francis Bacon*, vol. IV, London: Longman.

Banerjee, B. (2011) 'The Limitations of Geoengineering Governance in a World of Uncertainty', *Stanford Journal of Law, Science and Policy*, 4: 15–36.

Barnes, J. and Chin, J. (2018) 'The New Arms Race in AI', *The Wall Street Journal*, 2 March https://www.wsj.com/articles/the-new-arms-race-in-ai-1520009261

Barss, P. (2010) *The Erotic Engine*, Toronto: Anchor Canada.

Berlin, I. (1978 / 2013) 'The Decline of Utopian Ideas in the West', in *The Crooked Timber of Humanity*, 2nd edition, ed. H. Hardy, New York: Pimlico.

Berlin, I. (1983 / 2013) 'Giambattista Vico and Cultural History', in *The Crooked Timber of Humanity*, 2nd edition, ed. H. Hardy, New York: Pimlico.

Blackburn, S. (2005) *Truth*, London: Penguin.

Bostrom, N. (2002) 'Existential Risks: Analysing Human Extinction Scenarios and Related Hazards', *Journal of Evolution and Technology*, 9; page numbers from online version: https://www.nickbostrom.com/existential/risks.pdf

Briggle, A. (2005) 'Asilomar Conference', in *Encyclopedia of Science, Technology, and Ethics*, ed. C. Mitcham, Farmington Hills, MI: Macmillan.

Cam, P. (2012) *Teaching Ethics in Schools: A New Approach to Moral Education*, Camberwell, Australia: ACER Press.

Campion, N. (2008) *A History of Western Astrology, Volume 1, The Ancient World*, London: Bloomsbury.

Campion, N. (2009) *A History of Western Astrology, Volume 2, The Medieval and Modern Worlds*, London: Bloomsbury.

Caputo, J. (2013) *Truth: Philosophy in Transit*, London: Penguin.

Carlyle, T. (1841 / 2013) *On Heroes, Hero-Worship, and the Heroic in History*, eds. D. Sorensen and B. Kinser, New Haven, CT: Yale University Press.
Carrier, M. and Mittelstrass, J. (1991) *Mind, Brain, Behavior: The Mind-Body Problem and the Philosophy of Psychology*, Berlin: de Gruyter.
Chaudhuri, P. (2014) *The War with God: Theomachy in Roman Imperial Poetry*, Oxford: Oxford University Press.
Clarke, H. (2011) 'Steven Pinker and the An Lushan Revolt', *Quodlibeta* (website), 6 November https://bedejournal.blogspot.com/2011/11/steven-pinker-and-an-lushan-revolt.html
Claudian (395–7 / 1922) *Rape of Proserpine*, in *Claudian*, vol. II, trans. M. Platnauer, Cambridge, MA: Harvard University Press.
Comte, A. (1842 / 1974) *The Essential Comte*, ed. S. Andreski, trans. M. Clarke, London: Croom Helm.
Cox, B. and Cohen, A. (2017) *Forces of Nature*, London: Harper Collins.
Craig, C. (2015) '*Reform or revolution? Scott Sagan's* Limits of Safety *and its contemporary implications*', in *The Cuban Missile Crisis: A Critical Reappraisal*, eds. L. Scott and R.G. Hughes, Oxford: Routledge.
Crane, T. (2003) 'Mental Substances', in *Minds and Persons*, ed. A. O'Hear, Cambridge: Cambridge University Press.
Crane, T. (2014) 'Human Uniqueness and the Pursuit of Knowledge: a Naturalistic Approach', in *Contemporary Philosophical Naturalism*, eds. B. Bashour and H. Muller, London: Routledge.
Crane, T. (2017a) 'Philosophy, Science and the Value of Understanding', in *The Future of the Humanities in UK Universities: What is the Point of Philosophy?*, Stockholm: Ax:son Johnson Foundation.
Crane, T. (2017b) *The Meaning of Belief*, Cambridge, MA: Harvard University Press.
Crick, F. (1995) *The Astonishing Hypothesis: The Scientific Search for the Soul*, New York: Touchstone.
Dalai Lama XIV (1999) *Where Buddhism Meets Neuroscience: Conversations with the Dalai Lama on the Spiritual and Scientific Views of our Minds*, eds. Z. Houshmand, R. Livingston and B.A. Wallace, Boulder, CO: Shambhala Publications.
Davidson, D. (1986) 'A Coherence Theory of Truth and Knowledge', in *Truth and Interpretation*, ed. E. Lepore, Oxford: Basil Blackwell.
Dawkins, R. (2006a) *The God Delusion*, New York: Bantam.
Dawkins, R. (2006b) 'From the Afterword', letter to *The Herald* [UK newspaper], online at: https://www.heraldscotland.com/news/12760676.from-the-afterword/
Dawkins, R. (2011) *The Magic of Reality: How we know what's really true*, London: Transworld.
Democritus (fifth - fourth centuries BC / 2001) *Early Greek Philosophy*, ed. J. Barnes, 2nd edition, London: Penguin.
Dennett, D.C. (1973 / 1982) 'Mechanism and Responsibility', in *Free Will*, ed. G. Watson, Oxford: Oxford University Press.
Dennett, D.C. (1991) *Consciousness Explained*, Boston: Little, Brown and Company.
Dennett, D.C. (1992) 'The Self as a Center of Narrative Gravity', in *Self and Consciousness: Multiple Perspectives*, eds. F. Kessel, P. Cole and D. Johnson, Hillsdale, NJ: Erlbaum.
Dennett, D.C. (1995) '"The Mystery of Consciousness": An Exchange', *New York Review of Books*, 21 December 21 http://www.nybooks.com/articles/1995/12/21/the-mystery-of-consciousness-an-exchange/
Dennett, D.C. (2006) *Breaking the Spell: Religion as a Natural Phenomenon*, London: Penguin.

Dennett, D.C. (2012) 'The Mystery of David Chalmers', *Journal of Consciousness Studies*, 19: 86–95.
Dennett, D.C. (2016) 'Illusionism as the Obvious Default Theory of Consciousness', *Journal of Consciousness Studies*, 23: 65–72.
Dennett, D.C. (2017a) *From Bacteria to Bach and Back: The Evolution of Minds*, London: Penguin.
Dennett, D.C. (2017b) 'Interview with Carole Cadwalladr: I begrudge every hour I have to spend thinking about politics', *The Observer* [UK newspaper] https://www.theguardian.com/science/2017/feb/12/daniel-dennett-politics-bacteria-bach-back-dawkins-trump-interview
Derrida, J. (1967 / 1997) *Of Grammatology*, trans. G. C. Spivak, Baltimore, MD: John Hopkins University Press.
Descartes, R. (1642 / 1984) *Meditations on First Philosophy* and *Objections and Replies*, in *The Philosophical Writings of Descartes, vol. II*, trans. J. Cottingham, R. Stoothoff, and D. Murdoch, Cambridge: Cambridge University Press.
Desclos, M-L and Fortenbaugh, W. (eds.) (2011) *Strato of Lampsacus: Text, Translation, and Discussion*, New Brunswick, NJ: Transaction Publishers.
Diogenes of Oenoanda (second century / 2001) *Early Greek Philosophy*, ed. J. Barnes, 2nd edition, London: Penguin.
Douglas, A. (2012) 'Review of *The Atheist's Guide to Reality*, by Alex Rosenberg', *The Philosopher's Eye* http://wp.me/pcXlx-1pJ
Dretske, F. (1995) *Naturalizing the Mind*, Cambridge, MA: MIT Press.
Dunbar, K. and Fugelsang, J. (2005a) 'Scientific Thinking and Reasoning', in *The Cambridge Handbook of Thinking and Reasoning*, eds. K. Holyoak and R. Morrison, Cambridge: Cambridge University Press.
Dunbar, K. and Fugelsang, J. (2005b) 'Causal Thinking in Science: How Scientists and Students Interpret the Unexpected', in *Scientific and Technological Thinking*, eds. M. Gorman, R. Tweney, D. Gooding and A. Kincannon, Mahwah, NJ: Lawrence Erlbaum.
Durant, W. (1965) *The Story of Civilization, Volume 9: The Age of Voltaire*, New York: Simon and Schuster.
Eagleton, T. (2016) *Materialism*, New Haven, CT: Yale University Press.
Eddington, A. (1928 / 2012) *The Nature of the Physical World*, Cambridge: Cambridge University Press.
Epicurus (fourth–third centuries BC) *The Essential Epicurus*, trans. and ed. E. O'Connor, Amherst, NY: Prometheus Books.
Farrington, B. (1961) *Greek Science*, London: Pelican.
Feenberg (1999) *Questioning Technology*, Oxford: Routledge.
Feyerabend, P. (1963) 'Materialism and the Mind-Body Problem', *Review of Metaphysics*, 17: 49–66.
Feynman, R. (1965 / 2017) *The Character of Physical Law*, Cambridge, MA: MIT Press.
Field, H. (1972) 'Tarski's Theory of Truth', *Journal of Philosophy*, 69: 347–375.
Flanagan, O. (2011) *The Bodhisattva's Brain: Buddhism Naturalized*, Cambridge, MA: MIT Press.
Floris Cohen, H. (1994) *The Scientific Revolution: A Historiographical Inquiry*, Chicago: University of Chicago Press.
Foster, J. (1979) 'In Self-Defence', in *Perception and Identity: Essays Presented to A.J. Ayer*, ed. G.F. Macdonald, London: Macmillan.
Frankfurt, H. (2005) *On Bullshit*, Princeton, NJ: Princeton University Press.

Fukuyama, F. (2002) *Our Postmodern Future: Consequences of the Biotechnology Revolution*, London: Profile Books.
Galton, F. (1873) 'Africa for the Chinese', *The Times* [UK Newspaper], 5 June 1873 http://galton.org/letters/africa-for-chinese/AfricaForTheChinese.htm
Gaukroger, S. (2001) *Francis Bacon and the Transformation of Early-Modern Philosophy*, Cambridge: Cambridge University Press.
Geneva, A. (1995) *Astrology and the Seventeenth Century Mind: William Lilly and the Language of the Stars*, Manchester: Manchester University Press.
Goff, P. (2017) *Consciousness and Fundamental Reality*, Oxford: Oxford University Press.
Goff, P. (2019) *Galileo's Error: Foundations for a New Science of Consciousness*, London: Rider Press.
Gopnik, A. (2009) 'Could David Hume Have Known about Buddhism? Charles Francois Dolu, the Royal College of La Flèche, and the Global Jesuit Intellectual Network', *Hume Studies*, 35: 5–28.
Graves, R. (1955 / 1984) *Greek Myths*, Illustrated Edition, London: Penguin.
Gray, J. (2002) *Straw Dogs: Thoughts on Humans and Other Animals*, London: Granta.
Gray, J. (2004) *Heresies: Against Progress and Other Illusions*, London: Granta.
Gray, J. (2013a) 'The Limits of Materialism', Interview for BBC Radio 4, https://www.bbc.co.uk/programmes/b01s4vfs
Gray, J. (2013b) 'Myth Congeniality: John Gray Interviewed [by Nick Talbot]', *The Quietus* (website) http://thequietus.com/articles/12496-john-gray-silence-of-animals-interview
Gray, J. (2018a) *Seven Types of Atheism*, London: Allen Lane.
Gray, J. (2018b) 'Unenlightened Thinking: Steven Pinker's embarrassing new book is a feeble sermon for rattled liberals', *New Statesman*, 22 February https://www.newstatesman.com/culture/books/2018/02/unenlightened-thinking-steven-pinker-s-embarrassing-new-book-feeble-sermon
Gregory, F. (1977) *Scientific Materialism in Nineteenth Century Germany*, Dordrecht, Netherlands: Reidel.
Gross, N. (2008) *Richard Rorty: The Making of an American Philosopher*, Chicago: Chicago University Press.
Haack, S. (2016) *Scientism and its Discontents*, e-book published by *Rounded Globe*: https://roundedglobe.com/html/1b42f98a-13b1-4784-9054-f243cd49b809/en/Scientism%20and%20its%20Discontents/
Hand, M. and Winstanley, C. (eds.) (2008) *Philosophy in Schools*, London: Continuum.
Hargittai, I. (2010) *Judging Edward Teller: A Closer Look at One of the Most Influential Scientists of the Twentieth Century*, Amherst, NY: Prometheus Books.
Harris, S. (2012) *Free Will*, New York: Free Press.
Hawking, S. (1988) *A Brief History of Time*, New York: Bantam.
Hawking, S. and Mlodinow, L. (2010) *The Grand Design*, New York: Bantam.
Haynes, F. (ed.) (2016) *Philosophy in Schools*, London: Routledge.
Heidegger, M. (1927 / 1962) *Being and Time*, trans. J. Macquarrie and E. Robinson, Oxford: Basil Blackwell.
Henry, J. (2008) *The Scientific Revolution and the Origins of Modern Science*, 3rd edition, Basingstoke, UK: Palgrave Macmillan.
Hertzberg, A. (1968) *The French Enlightenment and the Jews: The Origins of Modern Anti-Semitism*, New York: Columbia University Press.
Hesiod (eighth century BC / 1993) *Works and Days* and *Theogony*, ed. R. Lamberton, trans. S. Lombardo, Indianapolis, IN: Hackett.

Hobbes, T. (1643 / 1976) *Thomas White's* De Mundo *Examined*, trans. H. Whitmore Jones, London: Bradford University Press.
Hobbes, T. (1654 / 1999) *Of Liberty and Necessity*, in *Hobbes and Bramhall on Liberty and Necessity*, ed. V. Chappell, Cambridge: Cambridge University Press.
Holbach, Baron d' (1770 / 1821) *The System of Nature*, Volumes I & II, trans. S. Wilkinson, Project Gutenberg e-book http://www.gutenberg.org/files/8909/8909-h/8909-h.htm [Volume I] / http://www.gutenberg.org/files/8910/8910-h/8910-h.htm [Volume II]
Horgan, J. (1999) *The Undiscovered Mind*, New York: Touchstone.
Hume, D. (1740 / 1975) *A Treatise of Human Nature*, ed. P. Nidditch, Oxford: Clarendon.
Ismael, J.T. (2016) *How Physics Makes Us Free*, Oxford: Oxford University Press.
Jackson, F. (1982) 'Epiphenomenal Qualia', *Philosophical Quarterly*, 32: 127–136.
Jacobson, N. (1969) 'The Possibility of Oriental Influence in Hume's Philosophy', *Philosophy East and West*, 19: 17–37.
Jacques, M. (2012) *When China Rules the World*, 2nd edition, London: Penguin.
James, W. (1907 / 1995) *Pragmatism*, Mineola, NY: Dover Publications.
Janaway, C. (1999) 'Will and Nature', in *The Cambridge Companion to Schopenhauer*, ed. C. Janaway, Cambridge: Cambridge University Press.
Jesseph, D. (2002) 'Hobbes's Atheism', *Midwest Studies in Philosophy*, 26: 140–166.
Johnstone, D. (2012) 'Review of *The Atheist's Guide to Reality*, by Alex Rosenberg', *The Independent*, 29 January http://www.independent.co.uk/artsentertainment/books/reviews/the-atheists-guide-to-reality-by-alex-rosenberg-6296076.html
Jonas, H. (1979 / 1984) *Imperative of Responsibility: In Search of an Ethics for the Technological Age*, Chicago: University of Chicago Press.
Kaku, M. (2008) *Physics of the Impossible*, London: Penguin.
Kant, I. (1787 / 1933) *Critique of Pure Reason*, trans. N. Kemp Smith, London: Macmillan.
Kaufman, D. (2016) 'Locke's Theory of Identity', in *A Companion to Locke*, ed. M. Stuart, Oxford: Blackwell.
Kelly, K. (2016) *The Inevitable: Understanding The 12 Technological Forces That Will Shape Our Future*, London: Penguin.
Kieckhefer, R. (1989) *Magic in the Middle Ages*, Cambridge: Cambridge University Press.
Kim, J. (2012) 'The very idea of token physicalism', in *New Perspectives on Type Identity: The Mental and the Physical*, eds. S. Gozzano and C. Hill, Cambridge: Cambridge University Press.
Kleeman, J. (2017) 'The race to build the world's first sex robot', *The Guardian*, 27 April https://www.theguardian.com/technology/2017/apr/27/race-to-build-world-first-sex-robot
Kohan, W. (2014) *Philosophy and Childhood: Critical Perspective and Affirmative Practices*, New York: Palgrave Macmillan.
Koons, R. and Bealer, G. (2010) 'Introduction', in *The Waning of Materialism*, Oxford: Oxford University Press.
Krauss, L. (2012a) *A Universe from Nothing: Why there is Something rather than Nothing*, London: Simon & Schuster.
Krauss, L. (2012b) 'The Consolation of Philosophy', *Scientific American*, 27 April https://www.scientificamerican.com/article/the-consolation-of-philos/
Kurzweil, R. (2005) *The Singularity is Near*, New York: Viking Penguin.
Ladyman, J. and Ross, D. (2007) *Every Thing Must Go: Metaphysics Naturalized*, with contributions from D. Spurrett and J. Collier, Oxford: Oxford University Press.
Landau, I. (1998) 'Feminist Criticisms of Metaphors in Bacon's Philosophy of Science', *Philosophy*, 73: 47–61.

Leary, T., Metzner, R., and Alpert, R. (1964 / 1990) *The Psychedelic Experience*, New York: Carol Publishing.
Lee, M. and Shlain, B. (1985) *Acid Dreams*, New York: Grove Weidenfeld.
Levine, J. (1983) 'Materialism and Qualia: The Explanatory Gap', *Pacific Philosophical Quarterly*, 64: 354–361.
Levine, P. (2017) *Eugenics: A Very Short Introduction*, Oxford: Oxford University Press.
Lipman, L. (1980) *Philosophy in the Classroom*, Philadelphia: Temple University Press.
Loar, B. (1997) 'Phenomenal States' in eds. N. Block, O. Flanagan and G. Güzeldere, *The Nature of Consciousness*, Cambridge, MA: MIT Press.
Locke, J. (1700 / 1979) *An Essay Concerning Human Understanding*, ed. P. Nidditch, Oxford: Clarendon Press.
Lucretius (first century BC / 2007) *The Nature of Things*, trans. A.E. Stallings, London: Penguin.
MacDonald, P. (2003) *History of the Concept of Mind, Volume 1*, Aldershot, UK: Ashgate.
Martinich, A.P. (1995) *A Hobbes Dictionary*, Oxford: Blackwell.
Matthews, G. (1980) *Philosophy and the Young Child*, Cambridge, MA: Harvard University Press.
Marx, K. (1841 / 2006) 'Difference Between the Democritean and Epicurean Philosophy of Nature', in *The First Writings of Karl Marx*, ed. P. Shafer, Brooklyn, NY: IG Publishing.
Marx, K. (1845 / 2000) 'French Materialism and the Origins of Socialism', in ed. D. McLellan, *Karl Marx: Selected Writings*, 2nd edition, Oxford: Oxford University Press.
Maxwell, N. (1984 / 2007) *From Knowledge to Wisdom: A Revolution for Science and the Humanities*, 2nd edition., London: Pentire Press.
McCall, C. (2009) *Transforming Thinking: Philosophical Inquiry in the Primary and Secondary Classroom*, London: Routledge.
McGinn, C. (1991) *The Problem of Consciousness*, Oxford: Blackwell.
McGinn, C. (2006) *Shakespeare's Philosophy: Discovering the Meaning Behind the Plays*, New York: HarperCollins.
McIntrye, L. (2018) *Post-Truth*, Cambridge, MA: MIT Press.
Merleau-Ponty, M. (1964 / 1968) *The Visible and the Invisible*, trans. A. Lingis, eds. A. Lingis and C. Lefort, Evanston, IL: Northwestern University Press.
Metzinger, T. (2009) *The Ego Tunnel: The Science of the Mind and the Myth of the Self*, New York: Basic Books.
Midgley, M. (2001) *Science and Poetry*, London: Routledge.
Mill, J.S. (1843 / 1965) *A System of Logic, Ratiocinative and Inductive*, London: Longmans, Green and Co. Ltd.
Monbiot, G. (2010) 'The Man who wants to Northern Rock the Planet', *The Guardian* [UK newspaper], 1 June https://www.monbiot.com/2010/06/01/the-man-who-wants-to-northern-rock-the-planet/
Montaigne, M. (1575 / 2003) 'Of the Education of Children', in *Michel de Montaigne: The Complete Works*, trans. D.M. Frame, London: Everyman.
Nagel, T. (1974) 'What is it like to be a bat?', *Philosophical Review*, 82: 435–450.
Naji, S. and Hashim, R. (eds.) (2017) *History, Theory and Practice of Philosophy for Children*, Oxford: Routledge.
Ney, A. (2008) 'Physicalism as an attitude', *Philosophical Studies*, 138: 1–15.
Nietzsche, F. (1883-8 / 1967) *The Will to Power*, ed. W. Kaufmann, trans. W. Kaufmann and R. Hollingdale, New York: Random House.
Nozick, R. (1974) *Anarchy, State, and Utopia*, Oxford: Blackwell.
O'Keefe, T. (2005) *Epicurus on Freedom*, Cambridge: Cambridge University Press.

Pacey, A. (1999) *Meaning in Technology*, Cambridge, MA: MIT Press.
Pandit, R. (2018) 'Pakistan remains ahead in nuclear warheads but India confident of its deterrence capability', *The Times of India*, 19 June https://timesofindia.indiatimes.com/india/pakistan-has-more-nuclear-warheads-india-credible-deterrence/articleshow/64641056.cms
Panek, R. (2011) *The 4 Percent Universe: Dark Matter, Dark Energy, and the Race to Discover the Rest of Reality*, Oxford: Oneworld.
Papineau, D. (2009) 'Physicalism and the Human Sciences', in C. Mantzavinos (ed.) *Philosophy of the Social Sciences: Philosophical Theory and Scientific Practice*, Cambridge: Cambridge University Press.
Parfit, D. (1984) *Reasons and Persons*, Oxford: Oxford University Press.
Persson, I. and Savulescu, J. (2012) *Unfit for the Future: The Need for Moral Enhancement*, Oxford: Oxford University Press.
Pinker, S. (1997) *How the Mind Works*, London: Penguin.
Pinker, S. (2006 / 2013) 'Deep Commonalities between Life and Mind', in *Language, Cognition, and Human Nature: Selected Articles*, Oxford: Oxford University Press.
Pinker, S. (2011) *The Better Angels of our Nature: A History of Violence and Humanity*, London: Penguin.
Pinker, S. (2018) *Enlightenment Now: The Case for Reason, Science, Humanism, and Progress*, New York: Penguin Random House.
Place, U.T. (1956) 'Is Consciousness a Brain Process?' *British Journal of Psychology*, 47: 44–50.
Place, U.T. (2002) 'A Pilgrim's Progress? From Mystical Experience to Biological Consciousness', *Journal of Consciousness Studies*, 9: 34–52.
Plato (fourth century BC / 1961) *The Collected Dialogues of Plato*, eds. E. Hamilton and H. Cairns, trans. L. Cooper, Princeton, NJ: Princeton University Press.
Plutarch (c. 98–125 / 1917) *Plutarch's Lives*, vol. V, trans. B. Perrin, Cambridge, MA: Harvard University Press.
Poliakov, L. (1968) *The History of Anti-Semitism, Volume III: From Voltaire to Wagner*, trans. M. Kochan, Philadelphia: University of Pennsylvania Press
Polkinghorne, J. (2002) *Quantum Theory: A Very Short Introduction*, Oxford: Oxford University Press.
Procopius (sixth century / 2007) *The Secret History*, trans. G.A. Williamson and P. Sarris, London: Penguin.
Putnam, H. (1983) 'Why there isn't a ready-made world', in *Realism and Reason, Philosophical Papers*, Volume 3, Cambridge: Cambridge University Press.
Quine, W.V.O. (1948 / 1961) 'On What There Is', in *From a Logical Point of View*, Cambridge, MA: Harvard University Press.
Quine, W.V.O. (1951 / 1961) 'Two Dogmas of Empiricism', in *From a Logical Point of View*, Cambridge, MA: Harvard University Press.
Quine, W.V.O. (1952 / 1976) 'On Mental Entities', in *The Ways of Paradox and Other Essays*, Cambridge, MA: Harvard University Press.
Quine, W.V.O. (1954 / 1957) 'The Scope and Language of Science', *The British Journal for the Philosophy of Science*, 8: 1–17.
Quine, W.V.O. (1957) 'Speaking of Objects', *Proceedings and Addresses of the American Philosophical Association*, 31: 5–22.
Quine, W. V. O. (1975) 'A Letter to Mr. Ostermann', in *The Owl of Minerva: Philosophers on Philosophy*, eds. C. Bontempo and S. Jack Odell. New York: McGraw-Hill.
Quine, W.V.O. (1985) *The Time of My Life: An Autobiography*, Cambridge, MA: MIT Press.

Quine, W.V.O. (1987) *Quiddities: An Intermittently Philosophical Dictionary*, Cambridge, MA: Harvard University Press.
Richardson, K. (2018) *Sex Robots*, Cambridge: Polity Press.
Ridley, M. (2015) *The Evolution of Everything: How New Ideas Emerge*, London: Fourth Estate.
Riesch, H. (2010) 'Simple or Simplistic? Scientists' Views on Occam's Razor', *Theoria*, 67: 75–90.
Roberts, P. (2012) 'Arkhipov, Vasili Alexandrovich (1926–1999)', in *Cuban Missile Crisis: The Essential Reference Guide*, ed. P. Roberts, Santa Barbara, CA: ABC-CLIO.
Rollins, S. (2014) 'The Real Sonny Rollins – In My Own Words', filmed interview with Bret Primack https://www.youtube.com/watch?v=aYt8B2RkqrM
Roochnik, D. (1996) *Of Art and Wisdom: Plato's Understanding of Techne*, University Park, PA: Pennsylvania State University Press.
Rorty, R. (1963 / 2014) 'Empiricism, Extensionalism, and Reductionism', in *Mind, Language, and Metaphilosophy*, eds. S. Leach and J. Tartaglia, Cambridge: Cambridge University Press.
Rorty, R. (1965 / 2014) 'Mind-Body Identity, Privacy, and Categories', in *Mind, Language, and Metaphilosophy*, eds. S. Leach and J. Tartaglia, Cambridge: Cambridge University Press.
Rorty, R. (1972 / 1982) 'The World Well Lost', in *Consequences of Pragmatism*, Minneapolis: University of Minnesota Press.
Rorty, R. (1979) *Philosophy and the Mirror of Nature*, Princeton, NJ: Princeton University Press.
Rorty, R. (1982) 'Introduction: Pragmatism and Philosophy', in *Consequences of Pragmatism*, Minneapolis: University of Minnesota Press.
Rorty, R. (1989) *Contingency, Irony, and Solidarity*, Cambridge: Cambridge University Press.
Rorty, R. (1994a / 1999) 'Religion as Conversation-stopper', in *Philosophy and Social Hope*, London: Penguin.
Rorty, R. (1994b / 1999) 'A World without Substances or Essences', in *Philosophy and Social Hope*, London: Penguin.
Rorty, R. (1998) *Truth and Progress: Philosophical Papers, Volume 3*, Cambridge: Cambridge University Press.
Rorty, R. (1999) *Philosophy and Social Hope*, London: Penguin.
Rorty, R. (2000) 'Response to Daniel Dennett', in *Rorty and his Critics*, ed. R. Brandom, Oxford: Blackwell.
Rorty, R. (2007 / 2010) 'Intellectual Autobiography', in *The Philosophy of Richard Rorty*, eds. R. Auxier and L. Hahn, Chicago: Chicago University Press.
Rossi, P. (1968) *Francis Bacon: From Magic to Science*, London: Routledge.
Rosenberg, A. (2011a / 2016) 'Why I Am a Naturalist', in *The Stone Reader*, eds. P. Catapano and S. Critchley, New York: Liveright.
Rosenberg, A. (2011b) *The Atheist's Guide to Reality*, New York: W.W. Norton.
Rosenthal, D. (2016) 'Quality Spaces, Relocation, and Grain', in *Sellars and his Legacy*, ed. J. O'Shea, Oxford: Oxford University Press.
Russell, B. (1925) 'Introduction: Materialism, Past and Present', in F. A. Lange, *The History of Materialism and Criticism of its Present Importance*, London: Routledge.
Ryan, J. (2010) *A History of the Internet and the Digital Future*, London: Reaktion.
Schopenhauer, A. (1859 / 1966) *The World as Will and Representation*, vol. 1, trans. E.F.J. Payne, New York: Dover Publications.

Sclove, R. (1995) *Democracy and Technology*, New York: Guilford Press.
Seigel, J. (2005) *The Idea of the Self*, Cambridge: Cambridge University Press.
Sellars, R.W. (1922) *Evolutionary Naturalism*, Chicago: Open Court.
Sellars, W. (1956 / 1997) *Empiricism and the Philosophy of Mind*, Cambridge, MA: Harvard University Press.
Sellars, W. (1981) 'Foundations for a Metaphysics of Pure Process', *The Monist*, 64: 3–90.
Shattuck, R. (1996) *Forbidden Knowledge: From Prometheus to Pornography*, New York: St. Martin's Press.
Shklovski, I. and Sagan, C. (1966) *Intelligent Life in the Universe*, San Francisco: Holden-Day.
Siderits, M. (2007) *Buddhism as Philosophy: An Introduction*, Farnham, UK: Ashgate.
Smart, J.J.C. (1959) 'Sensations and Brain Processes', *Philosophical Review*, 68: 141–156.
Smart, J.J.C. (1963a) *Philosophy and Scientific Realism*, London: Routledge.
Smart, J.J.C. (1963b) 'Materialism', *Journal of Philosophy*, 60: 651–662.
Smart, J.J.C. (2007) 'The Mind-Brain Identity Theory', in E. Zalta (ed.) *The Stanford Encyclopedia of Philosophy* https://plato.stanford.edu/archives/spr2017/entries/mind-identity/
Stern, D. (2007) 'Wittgenstein, the Vienna Circle, and Physicalism: A Reassessment', in A. Richardson and T. Uebel (eds.) *The Cambridge Companion to Logical Positivism*, Cambridge: Cambridge University Press.
Steward H. (1997) *The Ontology of Mind: Events, Processes, and States*, Oxford: Oxford University Press.
Steward, H. (2012) *A Metaphysics for Freedom*, Oxford: Oxford University Press.
Stoljar, D. (2010) *Physicalism*, London: Routledge.
Strawson, G. (1986) *Freedom and Belief*, Oxford: Clarendon Press.
Strawson, G. (2008) *Real Materialism and Other Essays*, Oxford: Oxford University Press.
Strawson, G. (2011) 'Radical Self-Awareness', in *Self, No-Self? Perspectives from Analytical, Phenomenological, and Indian Traditions*, eds. M. Siderits, E. Thompson and D. Zahavi, Oxford: Oxford University Press.
Swinburne, R. (1986) *The Evolution of the Soul*, Oxford: Clarendon Press.
Tallis, R. (1999) *Enemies of Hope: A Critique of Contemporary Pessimism*, Basingstoke, UK: Macmillan.
Tallis, R. (2011) *Aping Mankind: Neuromania, Darwinitis and the Misrepresentation of Humanity*, Durham, UK: Acumen Press.
Tallis, R. (2017) *Of Time and Lamentation: Reflections on Transience*, Newcastle upon Tyne, UK: Agenda Publishing.
Tartaglia, J. (2016a) *Philosophy in a Meaningless Life*, London: Bloomsbury.
Tartaglia, J. (2016b) 'Is philosophy all about the meaning of life?' *Metaphilosophy*, 47: 283–303.
Taylor, J. (2012) *Think Again: A Philosophical Approach to Teaching*, London: Continuum.
Teller, E. (1962) *The Legacy of Hiroshima*, Garden City, NY: Doubleday.
Teller, E. and Shoolery, J. (2001) *Memoirs: A Twentieth Century Journey in Science and Politics*, Cambridge, MA: Perseus Publishing.
Thagard, P. (2010) *The Brain and the Meaning of Life*, Princeton: Princeton University Press.
Tye, M. (1995) *Ten Problems of Consciousness*, Cambridge, MA: MIT Press.
Tyson, N. D. (2014) 'Neil deGrasse Tyson Returns Again', Podcast interview http://nerdist.nerdistind.libsynpro.com/neil-degrasse-tyson-returns-again.
Valberg, J.J. (1992) *The Puzzle of Experience*, Oxford: Oxford University Press.

Valberg, J.J. (2007) *Dream, Death, and the Self*, Princeton: Princeton University Press.
Valberg, J.J. (2011) *Will*, unpublished book.
Vickers, B. (ed.) (1984) *Occult and Scientific Mentalities in the Renaissance*, Cambridge: Cambridge University Press.
Weinberg, S. (1993) *Dreams of a Final Theory*, London: Vintage.
Williamson, T. (1995) 'Is Knowing a State of Mind?', *Mind*, 104: 533–565.
Williamson, T. (2011 / 2016) 'On Ducking Challenges to Naturalism', in *The Stone Reader*, eds. P. Catapano and S. Critchley, New York: Liveright.
Wilson, E.O. (2014a) 'Science, Not Philosophy, Will Explain the Meaning of Existence', *Big Think*, 4 November https://bigthink.com/think-tank/science-not-philosophy-will-explain-the-meaning-of-life-with-edward-o-wilson
Wilson, E.O. (2014b) *The Meaning of Human Existence*, New York: Liveright.
Winner, L. (1986) *The Whale and the Reactor: A Search for Limits in an Age of High Technology*, Chicago: University of Chicago Press.
Winter, T.N. (2007) 'The *Mechanical Problems* in the Corpus of Aristotle', University of Nebraska Faculty Publications, Classics and Religious Studies Department, Paper 68. http://digitalcommons.unl.edu/classicsfacpub/68
Wittgenstein, L. (1953) *Philosophical Investigations*, trans. G.E.M. Anscombe, Oxford: Blackwell.
Wittmann, B.C., Daw, N.D., Seymour, B. and Dolan, R.J. (2008) 'Striatal Activity Underlies Novelty-Based Choice in Humans', *Neuron*, 58: 967–973.
Wolf, S. (1997 / 2008) 'Meaning in Life', in *The Meaning of Life: A Reader*, eds. E.D. Klemke and S. Kahn, Oxford: Oxford University Press.
Wrenn, C.B. (2017) 'Truth is not (Very) Intrinsically Valuable', *Pacific Philosophical Quarterly*, 98: 108–128.
Wright, G. (2006) *Religion, Politics and Thomas Hobbes*, Dordrecht, Netherlands: Springer.
Young, F. (1964) *Giordano Bruno and the Hermetic Tradition*, London: Routledge.
Zirkle, C. (1941) 'Natural Selection before "The Origin of Species"', *Proceedings of the American Philosophical Association*, 84: 71–123.
Zuboff, S. (2019) *The Age of Surveillance Capitalism*, New York: Hachette.

Name Index

Agrippa, C. 24
Albert, D. 4
Ali, M. 136
Anscombe, E. 154
Aquinas 23
Archimedes 26
Aristotle 27, 90, 141–2
Arkhipov, V. 99–100
Armstrong, D.M. 184 n.16
Augustine 136
Austen, J. 176–7

Bacon, F. 24–5, 34, 183 n.26
Barthes, R. 10
Berkeley, G. 63
Berlin, I. 102
Blackburn, S. 195 n.11
Bostrom, N. 9, 27, 109
Bradley, F.H. 90
Buddha 149–50, 194 n.29
Butler, J. 145

Caesar 99, 168
Caputo, J.D. 163
Carlyle, T. 100, 102
Carnap, R. 36
Chaudhuri, P. 21
Churchland, P.S. 150
Claudian 20
Coleman, O. 156
Comte, A. 17
Conan Doyle, A. 88
Copleston, F. 83
Cox, B. 3
Crane, T. 154–6, 181 n.3
Crick, F. 61

Dalai Lama XIV 150
Darrow, C. 138–9
Darwin, C. 3, 6, 101–2, 116, 166
Davidson, D. 184 n.16

Dawkins, R. 3–6, 18, 23, 36, 103, 105
Deleuze, G. 37
Democritus 28–9, 33, 52, 69–70, 183 n.39, 184 n.7
Dennett, D.C. 36, 43, 60, 67–8, 99, 112–13, 121–2, 152, 169, 171
Derrida, J. 10, 163, 195 n.4
Descartes, R. 33, 49, 56–9, 75, 141–3, 186 n.2
Dunbar, K. 108
Durkheim, E. 10

Eagleton, T. 34, 183 n.3
Eddington, A. 33
Einstein, A. 4
Empedocles 6
Epicurus 33, 69, 119, 184 n.7

Farrington, B. 22–3, 182 n.19
Feyerabend, P. 61, 63
Feynman, R. 187 n.15
Ford, H. 58
Fox, J. 194 n.38
Frankfurt, H. 161
Freud, S. 10
Fugelsang, J. 108
Fukuyama, F. 113

Galileo, G. 34, 119, 136–7, 184 n.7
Galton, F. 102
Goff, P. 73, 184 n.7, 187 n.4
Gray, J. 104, 110–11, 113–17, 189 n.24, 190 n.51

Haack, S. 41
Harris, S. 126–8, 130–1
Hawking, S. 2, 6–8, 81, 104, 181 n.18
Heidegger, M. 183 n.25
Hesiod 19–20
Hempel, C. 35
Heraclitus 97

Hitler, A. 103, 113
Hobbes, T. 33–4, 119–20, 136–7, 184 n.4, 193 n.28
Holbach 101, 113, 116, 119, 136
Homer 104, 141, 189 n.30, 191 n.53
Hume, D. 133, 148–50

Ismael, J.T. 135–6

James, W. 163
Janaway, C. 192 n.12
Johnson, S. 63
Jonas, H. 9

Kant, I. 33, 73, 149–50, 154
Kelly, K. 97–8, 100, 112, 116
Kieckhefer, R. 23
Krauss, L. 3–6
Kripke, S. 36
Kurzweil, R. 103–4

Ladyman, J. 185 n.27
Landau, I. 183 n.26
Leach, S. 187 n.8
Leary, T. 151–2
Leucippus 183 n.38
Lewis, D.K. 184 n.16
Loar, B. 186 n.33
Locke, J. 133, 143–7, 152, 154–5, 193 n.5
Lucretius 11, 21, 25, 33–4, 98–9, 119, 148

McDowell, J. 36
McGinn, C. 65, 186 n.35
Mach, E. 50
McTaggart, J.M.E. 90
Malebranche, N. 58
Mallory, G. 107, 190 n.40
Marx, K. 33–5, 184 n.7
Maxwell, N. 105
Merleau-Ponty, M. 71–2
Metheny, P. 156
Metzinger, T. 194 n.29
Midgley, M. 33–4, 70
Mill, J.S. 133, 192 n.17
Molyneux, W. 144
Montaigne, M. 96, 110

Nagel, T. 36
Neurath, O. 53, 182 n.12

Ney, A. 35
Nietzsche, F. 6, 113, 162–6, 183 n.3
Nobel, A. 26
Nozick, R. 158–9

Ockham, W. 57

Papineau, D. 36
Paracelsus 24
Parfit, D. 36, 145–7, 150, 193 n.13
Persson, I. 105
Pinker, S. 4–5, 110–17, 183 n.29, 189 n.24, 190 n.50, 191 n.53, 191 n.59
Place, U.T. 51, 61–6
Plato 9, 20–2, 32–3, 48, 88, 96, 141–2, 150, 172
Procopius 136
Ptolemy, C. 136–7
Presley, E. 135
Price, H. 9
Putnam, H. 1, 36

Quine, W.V.O. 36, 50–7, 171

Ramsey, F.P. 163
Rees, M. 9
Ridley, M. 98–101, 116, 188 n.10, 188 n.12, 189 n.17
Rollins, S. 8
Rorty, R. 17, 56, 63, 163–6, 184 n.16, 186 n.29, 195 n.11
Rosenberg, A. 43–5, 51, 58, 185 n.27
Ross, D. 185 n.27
Russell, B. 36, 73, 83
Ryle, G. 51, 63

Sartre, J.-P. 73, 152
Savulescu, J. 105, 189 n.36
Schopenhauer, A. 111, 124, 126, 150, 192 n.12
Schrödinger, E. 41
Searle, J. 36
Sellars, R.W. 184 n.14
Sellars, W. 184 n.16
Shakespeare, W. 14, 16
Shelley, M. 11
Smart, J.J.C. 51–2, 56–8, 61–6, 133, 171, 186 n.5

Smollett, T. 186 n.8
Spencer, H. 101
Spinoza, B. 124
Steward, H. 132–3, 192 n.18
Stoljar, D. 51, 59
Strato 22–3
Strawson, G. 126, 151, 192 n.9
Strawson, P.F. 36
Swinburne, R. 149

Tallinn, J. 9
Tallis, R. 11, 40–1, 127–8, 187 n.11
Tarski, A. 163
Teller, E. 26
Tolstoy, L. 159

Trump, D. 113
Tyson, N.D. 2–3

Valberg, J.J. 73, 121, 155–6, 159
Voltaire 102

Watson, J. 51
Weinberg, S. 4
Williamson, T. 43, 167
Wilson, E.O. 5–6, 96
Winner, L. 9
Wittgenstein, L. 33, 45–6, 73, 161, 168, 182 n.12, 183 n.3
Wolf, S. 194 n.38

Xenophon 96

Lightning Source UK Ltd.
Milton Keynes UK
UKHW020230160522
403055UK00004B/100